BERM BREAKWATERS:

Unconventional Rubble-Mound Breakwaters

Derived from a Workshop at the
Hydraulics Laboratory, National Research Council
of Canada, Ottawa, Canada, September 15-16, 1987

Sponsored by the
Committee on Rubble Mound Structures
   of the Waterways, Port, Coastal and Ocean Division
   of the American Society of Civil Engineers
National Research Council of Canada
Canadian Society of Civil Engineers
American Shore and Beach Preservation Association

Edited by D. H. Willis, W. F. Baird and Orville T. Magoon

Published by the
American Society of Civil Engineers
345 East 47th Street
New York, New York 10017-2398

ABSTRACT

This book presents a state-of-the-art review of the design, analysis, hydraulic model testing, and prototype experience of a new class of coastal structures that produce economical designs using available quarry material under severe wave attack. Papers discuss how berm type breakwaters differ from conventional layered breakwaters Design procedures and parameters are reviewed. Prototype berm breakwaters are studied and research needs are identified.

Library of Congress Cataloging-in-Publication Data

Berm Breakwaters

    Includes indexes.

    1. Rubble mound breakwaters—Congresses. 2. Coastal Engineering—Congresses. I Willis, D.H. II Baird, W.F. III Magoon, Orville T. IV American Society of Civil Engineers Committee on Rubble Mound Structures.
    TC 33.B45 1988   627.24   88-16678
    ISBN 0-87262-663-6.

The Society is not responsible for any statements made or opinions expressed in its publications.

Authorization to photocopy material for internal or personal use under circumstances not falling within the fair use provisions of the Copyright Act is granted by ASCE to libraries and other users registered with the Copyright Clearance Center (CCC) Transactional Reporting Service, provided that the base fee of $1.00 per article plus $.15 per page is paid directly to CCC, 27 Congress Street, Salem, MA 01970. The identification for ASCE Books is 0-87262/88. $1 + .15. Requests for special permission or bulk copying should be addressed to Reprints/Permissions Department.

Copyright © 1988 by the American Society of Civil Engineers.
All Rights Reserved.
Library of Congress Catalog Card No.: 88-16678
ISBN 0-87262-663-6
Manufactured in the United States of America.

Cover photo: Hulguvik Bay Breakwater,
                 Keflavik, Iceland, by Bill Baird

FOREWORD

It is a pleasure to say a few introductory remarks today at this Seminar at the Canada Institute for Scientific and Technical Information, National Research Council of Canada, in Ottawa, Canada.

The purpose of this Seminar is to present individual ideas and synergistic discussions together. The conference has purposefully been kept small, because although you have thoughts and ideas on breakwaters you are present as experts and leaders in your respective fields such as coastal research, engineering, construction and design. I hope that you will discuss this exciting, evolving breakwater technology, for, although it draws much from old coastal engineering and hydraulic laboratory technology, it requires more study from the specialists here today so that it may be applied by coastal practitioners around the world.

We have several issues to resolve. What should we call this class of breakwaters? For the purposes of the next two days, I suggest that we use the names that we have been using individually, but in the discussions at the conclusion of our Seminar, let us settle on a name that can be used universally. My own feeling at this time is that "Berm Type Breakwater" is a good name.

Some of the other major topics to be discussed are:

- Are the berm type breakwaters separate and distinct from conventional layered breakwaters or are they a sub-class of layered breakwaters?
- How do the design procedures and design parameters differ from conventional breakwaters?
- What prototype experience has been obtained and how can it be promulgated in the future?
- What additional research is needed?

The idea for this conference was created during discussions in Ottawa between Bill Baird, Dave Willis and myself. We owe a great vote of thanks to Dave Willis, his staff and those at NRC as well as Bill Baird and the staff at W.F. Baird & Associates.

The material being presented today is both important and significant, and I am authorized as Chairman of the Rubble Mound Structures Committee of the American Society of Civil Engineers (ASCE) to publish a special volume containing all of your papers. Of course you may modify your papers or leave them as they are. I will correspond with each author shortly after the conference.

There will be a modest charge for this volume. A summary of this meeting will be presented by the ASCE Rubble Mound Structures Committee at the Coastal Engineering Conference in Spain, June 15-21, next year.

We are here as a group to meet challenges. We have seen the results of numerous failures both in terms of monetary losses and in terms of attacks on our profession. Part of our purpose is to bring together the many diverse areas of coastal expertise, with the background of many of the world's finest laboratories and world's best experts to publish our material. I believe the first step is to publish our proceedings, but we must also express cautions that this material is the result of the best available knowledge, but may not be ready for the general coastal engineering practitioner, without concurrent detailed model studies and analyses by experts.

I am delighted to be a part of this evolving coastal technology.

> Orville T. Magoon
> Chairman
> Rubble Mound Structure Committee
> American Society of Civil Engineers
> P.O. Box 279
> Middletown, CA 95461
> USA

PRÉAMBULE

Il me fait plaisir de formuler aujourd'hui quelques remarques d'introduction au présent séminaire qui se tient ici, à Ottawa, à l'Institut canadien de l'information scientifique et technique du Conseil national de recherches du Canada.

L'objet du présent séminaire est de permettre à chacun de présenter ses idées et d'en tirer ensemble des discussions synergiques. Le nombre de participants a été délibérément limité parce que, bien que vous ayez vos idées et conceptions propres en matière de brise-lames, vous êtes présents à titre d'experts et de chefs de file dans vos domaines respectifs, qu'il s'agisse de recherche sur les littoraux, ou encore de génie, de construction et de conception d'ouvrages côtiers. J'espère que vous débattrez de cette passionnante technologie en évolution qu'est celle des brise-lames et qui, bien que s'inspirant beaucoup d'anciennes techniques du génie des installations littorales et des travaux des laboratoires d'hydraulique, exige néanmoins une étude plus poussée de la part des spécialistes réunis ici, aujourd'hui, avant de pouvoir être appliquée par les practiciens des travaux littoraux de par le monde entier.

Nous devons solutionner plusieurs problèmes. Comment devrions-nous désigner cette classe de brise-lames? Au cours des deux prochains jours, je suggère que nous utilisions les noms par lesquels chacun d'entre nous les a individuellement désignés jusqu'ici, mais que, pendant les discussions de clôture de notre séminaire, nous convenions d'une désignation qui puisse être universelle. Pour ma part, je crois que l'expression "brise-lames de type risberme" convient bien.

Parmi les autres grands points à l'ordre du jour, mentionnons les suivants:

– Les brise-lames de type risberme constituent-ils une classe distincte de celle des brise-lames classiques à plusieurs couches ou une sous-catégorie des brise-lames à plusieurs couches?
– Comment les méthodes et paramètres de conception de ces brise-lames diffèrent-ils de ceux des brise-lames classiques?
– Quelle expérience a été acquise sur des prototypes et comment est-il possible de la disséminer à l'avenir?
– Quels aspects exigent des recherches plus poussées?

C'est lors de discussions tenues à Ottawa entre Bill Baird, Dave Willis et moi-même qu'est née l'idée de la présente conférence. Nous disons toute notre reconnaissance à Dave Willis, à son personnel et à celui du CNRC, ainsi qu'à Bill Baird et au personnel de la W.F. Baird & Associates.

La documentation présentée aujour'hui est d'une très grande importance et d'une très grande portée, et je suis autorisé, à titre de président du Comité sur les ouvrages en enrochement de l'American Society of Civil Engineers (ASCE) à publier un volume spécial renfermant toutes vos communications. Vous pourrez évidemment, selon votre choix, présenter une version intégrale ou modifiée de vos communications. Je communiquerai avec chacun des auteurs peu de temps après la présente conférence.

Ce volume sera vendu à un prix modique. Un résumé de la présente réunion sera présenté par le Comité sur les ouvrages en enrochement de l'ASCE à la Conférence sur le génie des installations littorales qui sera tenue en Espagne du 15 au 21 juin l'année prochaine.

Nous sommes ici présents en tant que groupe pour relever des défis. Nous avons tous constaté les résultats de nombreaux échecs, qu'il s'agisse de pertes monétaires ou d'atteintes à notre profession. Notre objectif est en partie de regrouper les nombreux et divers domaines d'expérience sur les littoraux, avec en toile de fond un grand nombre des travaux des meilleurs laboratoires au monde et al collaboration des meilleurs experts pour la publication de nos travaux. Je pense que la première étape consiste à publier un compte rendu, mais il faudra préciser que cette documentation, qui est basée sur les meilleures connaissances disponibles, peut ne pas encore être prête utilisable par l'ensemble des ingénieurs en installations littorales avant que des experts n'en fassent des études sur modèles et des analyses détaillées.

Je suis ravi de participer au développement de cette technologie des installations littorales.

    Orville T. Magoon
    Président
    Comité des ouvrages en enrochement
    American Society of Civil Engineers
    B.P. 279
    Middletown, CA 95461
    É.-U.

CONTENTS

CHAPTER

I.	Towards a Better Simulation of Sea States for Modelling of Coastal Structures E.P.D. Mansard	1
II.	Reef Breakwater Response to Wave Attack John P. Ahrens	21
III.	Rock Armouring to Unconventional Breakwaters: The Design Implications for Rock Durability N.W.H. Allsop and J.P. Latham	41
IV.	On the Stability of Berm Breakwater Roundheads and Trunk Erosion in Oblique Waves Hans F. Burcharth and Peter Frigaard	55
V.	Hydraulic Performance of Berm Breakwaters Ole Juul Jensen and Torben Sorenson	73
VI.	Application of Computational Model on Berm Breakwater Design J.W. van der Meer	92
VII.	Experimental and Historical Verification of the Performance of Naturally Armouring Breakwaters Kevin R. Hall	104
VIII.	The Development of a Design for a Breakwater at Keflavik, Iceland W.F. Baird and K. Woodrow	138
IX.	The Design and Construction of a Mass Armoured Breakwater at Hay Point, Australia W. Bremner, Dr. B.A. Harper and Prof. D.N. Foster	147
X.	Performance of a Berm Roundhead in the St. George Breakwater System Jeffrey F. Gilman	219
XI.	Implementation and Performance of Berm Breakwater Design at Racine, WI Robert J. Montgomery, Gregory J. Hofmeister and William F. Baird	229

XII.	Berm Type Armor Protection for a Runway Extension at Unalaska, Alaska *Charles I. Rauw*	250
XIII.	Unconventional Rubble-Mound Breakwaters—Concerns *C.D. Anglin, K.B. Dean and D.H. Willis*	270

List of Participants 279

Subject Index 283

Author Index 284

TOWARDS A BETTER SIMULATION OF SEA STATES
FOR MODELLING OF COASTAL STRUCTURES

by

E. P. D. Mansard

Abstract

In the last fifteen years the techniques of wave generation have advanced to a great extent that it is now possible to exercise controls, on parameters such as wave grouping, wave asymmetries, etc., during the simulation process. This paper describes some of these techniques that were developed at the National Research Council of Canada, towards a better understanding of Nature and its reproduction. Emphasis is given to the non-linearities associated with shallow water waves and the difficulties encountered in depth limited situations due to complex shoaling mechanism.

Résumé

Au cours des 15 dernières années les méthodes de génération des vagues ont tellement progressé qu'il est maintenant possbile de contrôler des paramètres comme le groupement des vagues, les asymétries des vagues, etc., pendant le processus de simulation. Cette étude décrit certaines des méthodes mises au point au Conseil national de recherches du Canada pour une meilleure compréhension de la nature et sa reproduction plus fidèle. L'emphase a été placée sur les non-linéarités associées aux vagues en eaux peu profondes et sur les difficultés que pose le complexe mécanisme de la diminution des profondeurs en situations de profondeur restreinte.

TOWARDS A BETTER SIMULATION OF SEA STATES
FOR MODELLING OF COASTAL STRUCTURES

by

E.P.D. Mansard
Hydraulics Laboratory
National Research Council of Canada
Ottawa, Ontario, K1A 0R6
Canada

ABSTRACT

In the last fifteen years, techniques of wave generation have advanced to such an extent that it is now possible to exercise controls of parameters such as wave grouping, wave asymmetries, etc., during the simulation process. This paper describes some of the techniques that were developed at the National Research Council of Canada, to enable a better understanding of nature and its reproduction in model basins. Emphasis is given to the non-linearities associated with shallow water waves and the difficulties encountered in depth limited situations due to complex shoaling mechanisms. These findings have important consequences in the correct modelling of coastal structures and moored vessels in the nearshore zone.

1.0 INTRODUCTION

Physical modelling is a technique which is commonly used in the design of coastal structures. One of the key elements in this design procedure is the simulation of the sea states. In the last fifteen years, the techniques of wave simulation have evolved rather rapidly, from generation of regular waves, to irregular waves, and now to generation of multi-directional waves; but a thorough understanding of naturally occurring waves has not yet been achieved. Most laboratories attempt to subject their test structures to 'realistic' sea state inputs; but the precise definition of 'realistic' is still a subject of intense research. Nevertheless, a variety of sophisticated generation techniques are now available to the researcher so that he can exercise control over wave profiles, wave height statistics, groupiness properties, etc., in his studies.

The Hydraulics Laboratory of the National Research Council of Canada has been actively involved since 1969 in developing methods for generation and analysis of waves. This paper traces the evolution of the generation techniques over the years and is intended to share some of NRC's experience on the relevance of these tools in terms of structural response.

Recently NRC has developed the capability to subject test structures to multi-directional wave attacks (Miles et al., 1986; Miles and Funke, 1987; Funke and Miles, 1987), but this paper will restrict itself to the developments achieved for uni-directional seas.

2.0 WAVE SYNTHESIS TECHNIQUES

In model basins and flumes, wave generation is achieved in two steps. First, a time series of specified record length, representing the water surface elevation of the desired sea state is synthesized. Proper compensations are then applied to convert this time series into a voltage signal which can drive the wave paddle and ensure the desired simulation[1].

A sea state is generally expressed in terms of its variance spectral density. There exists a variety of techniques by which the water surface elevation can be synthesized from this spectral density. These different techniques may be classified by three different approaches:

- non-deterministic approach;
- partly deterministic approach; and
- deterministic approach.

The non-deterministic approach is one which produces a non-repeating wave train. Methods that draw their random noise sources directly from thermal or cosmic sources belong to this category. In a laboratory environment, this method is often unsuitable because of its non-repeatability.

In principle, all other methods of synthesis which do not fall into the above category could be classified as being deterministic, since the user is called upon to make certain choices. However, the concept of deterministic simulation has usually been closely linked only to the laboratory reproduction of given prototype wave trains. All other methods to produce wave trains based on certain choices could be classified as being partly deterministic. An extensive review of the various techniques may be found in Funke and Mansard (1987).

Table 1 gives a summary of the different synthesis techniques presently used by NRC. As can be seen from this table, most techniques belong to the category of partly deterministic. The choice of a specific technique from those listed in Table 1 depends on the requirements of the study.

Recent investigations have led to development of new techniques which can simulate some of the non-linearities associated with waves. Table 1 lists those developments that have been implemented at NRC.

---

[1] A number of analysis algorithms are obviously required to verify the effectiveness of the simulation. A discussion on these algorithms is beyond the scope of this paper.

> **DETERMINISTIC APPROACH**
>
> Reproduction of scaled prototype wave train
>
> **PARTLY DETERMINISTIC APPROACH**
>
> **NO SPECTRAL OR TIME DOMAIN CONTROL BUT REPEATABLE SIMULATION**
>
> Fourier synthesis using filtered output of random complex spectrum
>
> **SPECTRAL DOMAIN CONTROL (REPEATABLE)**
>
> Fourier synthesis with random phase spectrum
>
> **SPECTRAL AND TIME DOMAIN CONTROLS (REPEATABLE)**
>
> Fourier synthesis with random phase spectrum using a preselected random number seed
>
> Grouped wave train using the concept of 'SIWEH'
>
> **GENERATION OF EPISODIC WAVES AND TRANSIENTS (REPEATABLE)**
>
> **CONTROL OF NON-LINEARITIES**
>
> Distortion of wave profiles
>
> Correct reproduction of sub-and super-harmonics

TABLE 1

AN OVERVIEW OF WAVE GENERATION TECHNIQUES AT NRC

2.1 Deterministic Approach

2.1.1 Reproduction of scaled prototype wave train

Since the complexity of nature is not well understood, some laboratories strongly advocate the use of prototype wave trains for model studies. This ensures that both the frequency as well as the time domain characteristics of nature are well reproduced[1]. The obvious limitation to this approach is the lack of wave records for all sites of interest or for the worst storm conditions that may prevail. Canada has a wealth of wave records for its coasts and inland waters. Since 1970, more than 150 stations have, at one time or other, moni-

---

[1] Although this method could be considered as being deterministic in a 2D-situation, the directionality of the wave components have to be reproduced in order for this to be rigorously deterministic.

tored the wave climate. The Hydraulics Laboratory of NRC has used some of these wave records, particularly those from the Hibernia field (off the Newfoundland coast) in model studies. Many of these records have displayed distinct wave grouping.

2.2 Partly Deterministic Approach

2.2.1 Fourier synthesis using filtered output of random complex spectrum

This method, often known as the random complex spectrum method, consists of generating a time series of water surface elevation of specified record length, from a white noise spectrum. A Gaussian distributed white noise complex spectrum with a standard deviation of 1 is first generated and then filtered using the desired amplitude spectrum. Subsequent inverse Fourier transform results in the desired time series. For more details on this method, the reader is referred to Funke and Mansard (1984, 1987). Although this technique could fall into the category of the non-deterministic approach, it is classified here under the partly-deterministic group because of the constraint that the record length is finite. Furthermore, the same wave train can be reproduced over and over again as long as the random number generator is the same. The wave train synthesized by this method will not have the specified spectral density at every single realization since it will only represent one possible sample of an infinitely long process. However, when averaged over a number of realizations, the synthesized spectrum will match the specified spectral density.

2.2.2 Fourier synthesis with random phase spectrum

This method, often referred to as the random phase spectrum method, is probably the most commonly used. This technique is also based on the inverse fast Fourier transform; it pairs a given target spectrum with a randomly selected phase spectrum. Unlike the previous method, this technique reproduces the target spectrum at every single realization no matter how short the record length. In the classification of the partly deterministic approach, this method exercises "spectral determinism" by reproducing the frequency domain characteristics of a given sea state. However, no specific controls are exercised over the time domain characteristics. These time domain characteristics will vary, both in this as well as in the previous method, according to the sequence of random numbers used in the simulation.

Each of the above two methods have their own proponents who justify their use. For instance, Tucker et al. (1984) claimed that the simulation based on the random phase spectrum method incorrectly reproduces the ocean surface, and therefore that some of the wave parameters such as wave grouping were likely to be in error. It seems, however, that this claim is not entirely supported by others (Mansard and Funke, 1984; Elgar et al., 1985; Medina and Aguilar, 1985). For time series whose recycling length becomes very long, both the random phase spectrum method of synthesis and the random complex spectrum method will result in similar wave parameters. However, if the record length is short, a certain variability will be found in the results of both these methods. In fact, it was found by Mansard and Funke (1984) that there is as large a difference between two records selected at random from one method than there is between the two methods. Since the impact of this conclusion is far-reaching in model tests, some of the highlights of the work caried out at NRC in this regard are given below.

2.2.3 Comparison between random complex spectrum and random phase spectrum methods

Using the random phase spectrum method, Mansard and Funke (1984) synthesized 200 time series, each with different time domain characteristics, from a given spectrum. This was achieved by varying the first seed in the random number generator each time the program was recycled for the generation of a new time series. Different seeds result in different sequences of random phases and when incorporated in the inverse Fourier Transform, they resulted in varying time domain features. These time series, each with a duration of 200 s, were then subjected to several frequency and time domain analyses. More than twelve parameters were computed and then assembled into independent wave parameter time series for statistical analysis.

This entire analysis sequence was then repeated, using this time the random complex spectrum method of synthesis instead of the random phase spectrum method. Figure 1 illustrates some of the interesting results of these investigations. As can be seen, the JONSWAP peak enhancement factor γ was changed from 1 to 3.3 to 7 in order to evaluate the influence of spectral width on the wave parameters. In the original work (Mansard and Funke, 1984) the variance of the spectrum was allowed to increase with γ; but for these tests, the different wave parameters were recomputed, this time keeping the variance constant.

Inspection of Figure 1, which shows the mean value and the range between maximum and minimum values of some selected parameters, leads to the following conclusions:

- The differences between the random phase and the random complex spectrum methods for wave synthesis, in terms of several major parameters, are small.

- The values of peak frequency correspond to those desired using the Delft method (List of Sea State Parameters, 1986). This method computes the centroid of the spectral band between the lower and upper intercepts of the spectral density, and the threshold, which is 80% of the maximal spectral value. The mean values of the peak frequency, resulting either from the random phase spectrum or the random complex spectrum method are very close to the expected values. However, the random complex spectrum method has larger standard deviations. The variations found in the random phase spectrum method are principally due to the use of the Welch method for spectral analysis.

- There is a correlation between the spectral width and the mean value of groupiness factor (GF). This implies that

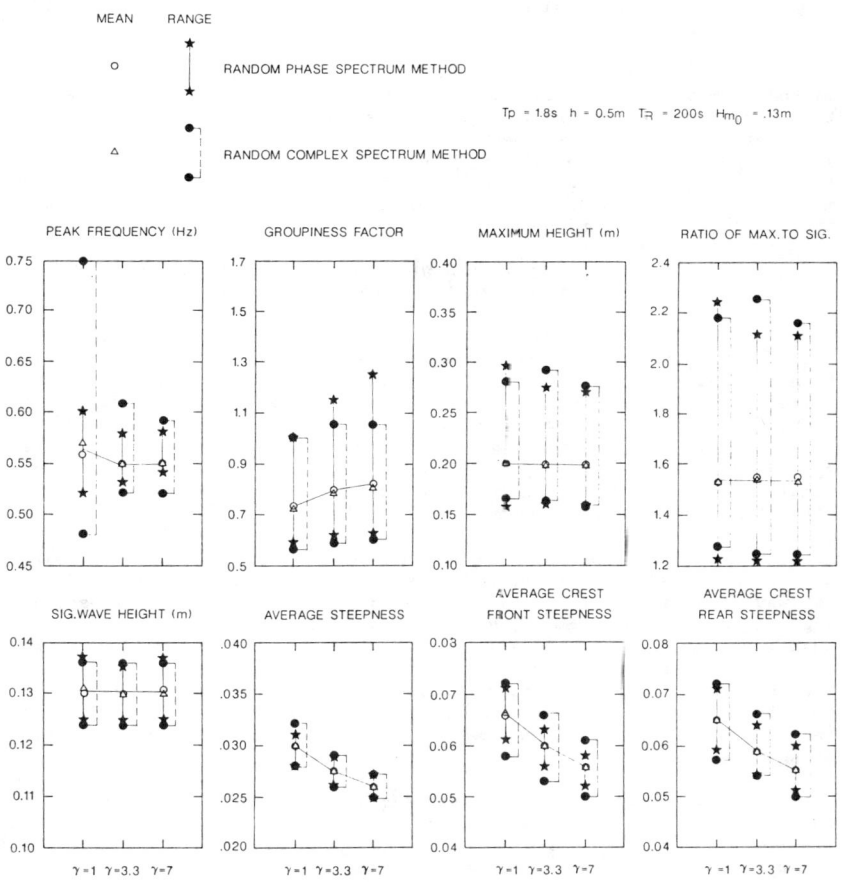

FIGURE 1

Statistical variability of selected wave parameters

- narrower spectra can result in higher GF if the record is sufficiently long. For shorter records, however, it is generally possible to achieve a given value of GF either by a narrow or a broad spectrum.

- The average value of the various steepnesses is higher with broader spectra.

Since Tucker et al.'s claim, that the random complex spectrum method provides more realistic simulation, has not been substantiated, the Hydraulics Laboratory of NRC prefers using the random phase spectrum method. Often, the client requires the spectral domain characteristics - like the shape of the spectrum and the peak frequency, etc. - be well reproduced. For these requirements, the random phase spectrum method is superior since these criteria can be achieved without resorting to longer testing times.

2.2.4 Fourier synthesis with random phase spectrum method using a preselected random number seed

In the past, there have been cases when NRC was asked to use the random phase spectrum method and, at the same time, to provide certain specific controls on the time domain characteristics of the sea state. For instance, one client wanted the ratio of $H_{max}/H_{1/3}$ (RMOSIG) to be equal to 1.86 within the specified record length. Alternatively, some situations might warrant the simulation to have specified values of this RMOSIG and also a certain groupiness factor, simultaneously. To satisfy these requirements, a new synthesis algorithm was developed. Details of this technique are as follows.

Given a spectral density, a number of wave trains each with different time domain characteristics were synthesized. These different wave trains were then analysed independently for their corresponding values of GF and RMOSIG, keeping track, however, of the seed value that was used in each realization. Subsequently, a relationship was obtained in a tabular form, between the seed value used in the synthesis and the resulting parameters. Finally, an algorithm was developed to perform a table look-up in order to select the seed corresponding to the client's request of parameters and to synthesize a wave record using that seed.

Obviously this method has limitations since the relationship between the seeds and the wave parameters can change as a function of the spectrum and the record length. To cover all possible situations an extensive library of these tables would be required. However, this technique was developed to illustrate the point that one could explicitly determine the required seed value by numerical simulation and then reproduce the desired parameters in the wave flumes.

2.2.5 Grouped wave train using the concept of 'SIWEH'

One of the time domain characteristics which has gained a lot of importance in recent years is the phenomenon of wave grouping. In order to exercise control on the degree of grouping, a synthesis technique was developed at NRC using the concept of 'SIWEH'. (Funke and

Mansard, 1979) This technique basically consists of pairing, by means of the inverse FFT, the amplitude spectrum provided by the variance spectral density and the phase spectrum computed from the SIWEH (<u>S</u>moothed <u>I</u>nstantaneous <u>W</u>ave <u>E</u>nergy <u>H</u>istory). The SIWEH, which represents the smoothed function of the square of the water surface elevation, can be derived either from prototype records or synthesized from a model SIWEH spectral density. In the absence of any suitable model for the above spectral density, NRC uses a model proposed by Funke and Mansard (1979) which gives control over the period of the groups and the groupiness factor. Recently Pinkster (1984) has given a formulation for the groupiness spectral density, based on a random Gaussian stochastic process. However, this function is applicable only for very long records. For shorter realizations, there may be appreciable deviations from the expected curve.

It has been found that some of the wave trains, particularly those recorded in the Hibernia field, display distinct grouping. The SIWEH obtained from one of these records was used to illustrate the importance of wave grouping on breakwater stability (Johnson et al., 1978). More recently, Mansard and Pratte (1982) used the SIWEH concept in a model study of a moored tanker. By synthesizing three wave trains, all having the same variance spectral density, but different groupiness factors, a strong dependency between the horizontal oscillations of the tanker and the degree of grouping was shown.

2.2.6 Generation of episodic waves and transients

In order to determine extreme responses from test structures, the Hydraulics Laboratory of NRC has developed a technique to simulate episodic waves which can break exactly at the test structure.

In the past, the generation of these kinds of breaking waves were realized by several somewhat arbitrary sweep frequency techniques. The technique which is used by NRC is unique in the sense that it can exercise easy control on the desired time function of the breaking wave. No attempt is made, however, to match any spectral shape. An illustration of this technique is given in Figure 2. Figure 2a shows an episodic wave breaking on a conical structure while Figure 2b corresponds to wave slamming on a berthing dolphin. For further details on this technique, the reader is referred to Mansard and Funke (1982).

2.3 CONTROL of NON-LINEARITIES

2.3.1 Distortion of wave profiles

Most random wave simulation techniques assume that wind generated waves can be satisfactorily represented within the framework of the Gaussian hypothesis. However, several parameters, such as wave asymmetries, are known to depart from this hypothesis. If waves are truly Gaussian, the value of horizontal asymmetry (ratio of crest height to total wave height) should be equal to 0.5, while the vertical asymmetry (ratio of crest front steepness to crest rear steepness) should correspond to a value of 1. Myrhaug and Kjeldsen (1984) who

FIGURE 2

Simulation of episodic waves.

carried out extensive work on these parameters have illustrated that prototype waves are asymmetric, even in deep water.

None of the classical wave generation techniques can simulate these asymmetries in a controlled fashion. In the past, these asymmetries were realized, to a certain extent, in wind wave flume facilities, where the wind was used to steepen the waves. Recently, the Hydraulics Laboratory developed a technique to control asymmetries through a set of non-linear transformations. These transformations can impose a certain degree of asymmetry in the wave profiles keeping, however, the spectral density the same. Detailed information of these transformations and the laboratory reproduction of these asymmetries can be found in Funke and Mansard (1982).

Although changes in asymmetries can be realized by these techniques, it is not known at this time what asymmetries ought to be used to achieve a good match with nature. Waves recorded by Waverider buoys do not exhibit a correct degree of asymmetry. This is due to the fact that non-linear mooring characteristics of the buoy in association with the local particle velocity of the waves tend to distort the profile. Following some recent investigations on 2nd order waves, the horizontal asymmetries can be explained to a large extent by the presence of super-harmonics. A description of these super-harmonics is given in the next section.

2.3.2 Correct reproduction of sub- and super-harmonics

Simulation of waves in numerical and physical models is often done using first order linear wave theory. However, if the Laplace equation describing the waves is solved to second order, two important phenomena occur. These are called sub- and super-harmonics.

Using the concept of radiation stress, it can be shown that a set-down of mean water level (MWL) occurs under the groups, while in between the groups, a corresponding set-up occurs. This results in a long wave oscillation known as sub-harmonics (or bounded long waves) with a period equal to the interval between groups. The sub-harmonics are bound to the group structure and travel with group velocity. On the other hand, super-harmonics (or bounded short waves) are part of the double frequency phenomenon which distorts the wave profile by sharpening the crest and flattening the trough.

The transfer functions which relate the magnitudes of these second order waves to their primary counterparts are illustrated in Barthel et al. (1983) for sub-harmonics, and in Sand and Mansard (1986) for super-harmonics. Although these magnitudes can be computed relatively easily and taken into account in numerical models, their reproduction in physical models is not straightforward. Because of the bounded character of these second order harmonics, they cannot be reproduced correctly by the first order linear wave generation technique.

When a first order wave train is generated using a linear wave generation technique, the bounded harmonics (both sub- and super-) are automatically created. However, since the boundary conditions at the waveboard are not satisfied properly up to second order, a number of spurious components are generated. These spurious components then propagate along the flume as free waves with celerities corresponding to the dispersion relation, causing a certain interference with the bounded waves.

In cooperation with the Danish Hydraulic Institute, and the Delft Hydraulics Laboratory, the National Research Council of Canada, solved the Laplace equation and the wave board boundary conditions to second order and found expressions relating the spurious components to the sub-harmonics. Proper compensation was then applied during the generation process to eliminate these spurious (or parasitic) components by generating waves with amplitudes the same as spurious waves, but of opposite phases. By performing this operation, the spurious waves were eliminated leaving behind only their bounded counterparts. The effectiveness of this technique was extensively tested and interested readers should refer to Barthel et al. (1983) for more details.

It was clear from these experiments that if spurious waves were not eliminated, a correct set-down under the groups was no longer achievable. In addition, the energy content associated with the long waves was incorrect.

More recently, a similar cooperation with the Danish Hydraulic Institute led to the control of super-harmonics. Details of the theoretical investigations and illustration of experimental realizations can be found in Sand and Mansard (1986). It should be pointed out that, although these latter techniques are somewhat similar to those given for sub-harmonics, additional terms dealing with twice the fundamental frequencies had to be included in this new technique.

Recently Barthel and Mansard (1987) extended the experimental validation techniques, which were earlier carried out only for piston mode, for waveboards operating in flap and combined modes.

Implications on Coastal Studies

The general water surface boundary condition is particularly non-linear in shallow water. Therefore, the magnitudes of both super- and sub-harmonics increase as the water depth decreases. This means that, when testing coastal structures, these harmonics should be correctly reproduced in order to obtain realistic responses from the structures.

As an example, Mansard and Pratte (1982) found that the correct reproduction of the sub-harmonics led to an increased response of a 227,000 dwt model vessel in shallow water, when compared to the results obtained using classical first order wave generation theory. The horizontal oscillations, which are affected by the sub-harmonics, increased in magnitude when proper set-down under the groups was ensured. This resulted in increased loads on all the mooring lines. Similar controls over harmonics used to study wave run-up on a model beach showed that the statistics of run-up parameters are over-estimated if proper compensation for spurious long waves was not applied (Barthel et al., 1983). Pedersen et al. (1987) have also recently established the necessity for this compensation while investigating the loads on a tanker moored behind an offshore terminal.

In terms of structural response due to super-harmonics, experimental investigations are underway; but recently Mansard et al. (1986) analysed prototype wave records from three ranges of water depths and showed that the presence of super-harmonics can result in higher crest heights and horizontal velocities. These of course lead to increased loading on and overtopping of fixed structures.

It is anticipated that, while model testing coastal structures, the response could be different depending upon the location of the test structure from the paddle. The super-harmonics created by the low frequency part of the 1st-order (primary) spectrum have overlapping frequency domains, with the primary components present in the high frequency tail. Due to their different celerities, (ie. super-harmonics travelling with bound velocity and primary component following the linear dispersion relation) substantial variations in the high frequency part of the spectrum should be found along the path of propagation. By numerical simulations Sand and Mansard (1986) were able to illustrate that these differences could be substantial as the depth of water decreases. They also showed that these oscillations in the high frequency tail result in varying crest front steepnesses along the flume. Hence, in a model study of a coastal structure, it is possible that the structural response is different depending upon the location of the structure.

3.0 GENERATION OF DRIVING SIGNAL

In previous sections, a brief review of the various synthesis techniques was presented. These techniques produce basically a time series of water surface elevation with desired characteristics. This time series then has to be compensated for wave machine characteristics before its correct reproduction can be achieved in a model set-up. Furthermore, when a particular wave train has to be generated it is necessary that this be achieved at the test section, which may be 20 to 30 m away from the paddle. Therefore, suitable compensations have to be applied and they constitute, by themselves, an important element in the overall generation package.

The program RWREP used by NRC for this purpose, performs a variety of compensations, and also permits the user to impose certain controls to protect the wave machinery from excessive strokes and velocities. A summary of its main features is given in Table 2, and an example of its performance is illustrated in Figure 3. Full details on this program and its capabilities may be found in Funke and Mansard (1984).

4.0 Effects of shoaling on wave statistics

Most of the techniques described above are applicable principally to constant depth situations. In model studies which simulate depth limited situations, where shoaling of waves is predominant, the various controls associated with these techniques are relevant only to the constant depth portion of the model. For convenience, this region

> Main Steps in the Program RWREP
>
> 1. Fourier transform of data.
>
> 2. Phase compensation for reproduction at the test structure.
>
> 3. Band pass filtering for restricting excessive stroke and acceleration.
>
> 4. Complex compensations for:
>
>   - Dynamic relationship of water surface variation to corresponding wave board position.
>
>   - Relationship between the actuator and wave board position.
>
>   - Servo system response.
>
>   - Dynamic characteristics of the analog low pass filter which follows the digital to analog converter.
>
> 5. Inverse Fourier transform.
>
> 6. Application of the following safety measures:
>
>   - Slew rate limit for excessive velocities.
>
>   - Recycling the data to avoid start-up transients and clipping of the signal to the maximum allowed voltage. (This is based on the voltage corresponding to the maximum stroke of the machine.)

TABLE 2

CONVERSION OF A WAVE TRAIN INTO A WAVE BOARD COMMAND SIGNAL

will be referred to as "offshore" in this paper while the depth limited region will be called as "nearshore". When shoaling takes place in the nearshore region, the wave characteristics undergo certain transformations which are not well understood. Accordingly, these transformations are not accounted for in the different techniques presented above. It is customary, therefore, to simulate a pre-specified sea state in the offshore region of the model, and to let the shoaling take its natural course on a bathymetry similar to the one found in nature. The final design of the structure is then related to the corresponding offshore wave conditions and their probabilities of occurrence. Recently, however, investigations have shown that for a given offshore sea state, the severity of the wave climate realized in the nearshore could be different depending upon the time domain characteristics of that offshore sea state. For instance, it can be postulated that a highly grouped wave train, undergoing shoaling may have its larger waves become unstable and break before they reach the struc-

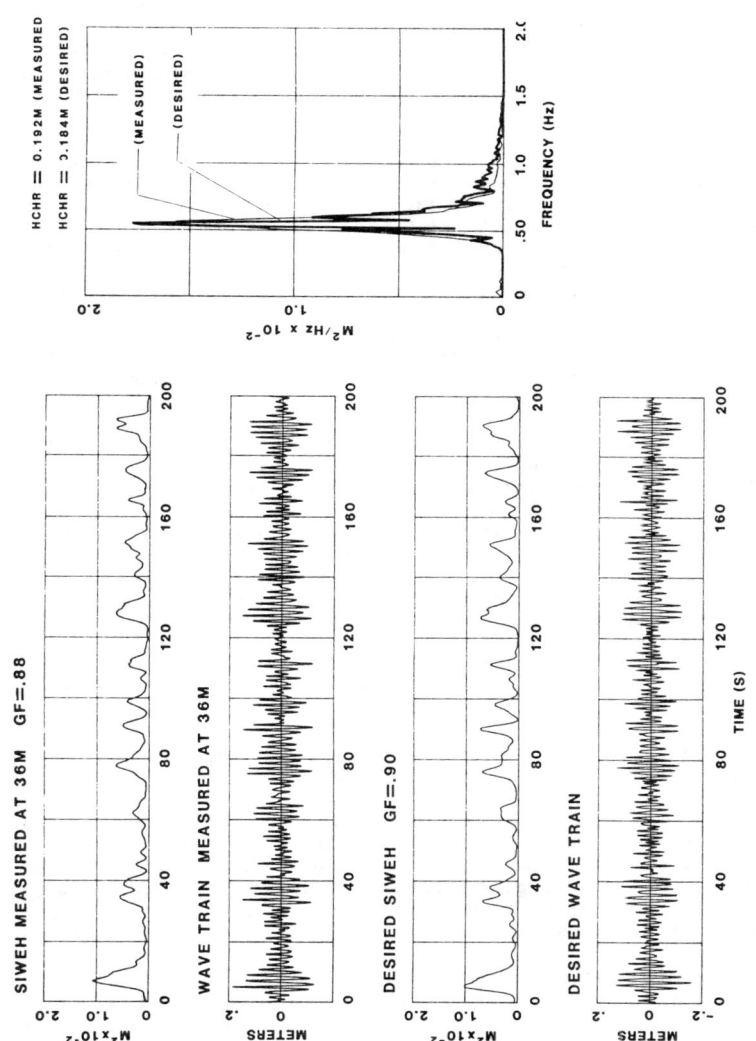

FIGURE 3

Reproduction of the 1st order wave characteristics at the test section

ture. On the other hand, with a lower degree of grouping, the waves might just shoal to their maxima and impact on the structure, causing increased damage. This hypothesis has been proven correct in some cases. But the underlying wave transformation is much more complex and varies for each combination of period and height of the waves, bathymetry, groupiness factors, etc. Furthermore, the non-linear components of the waves are found to increase with decreasing depths and complicate the entire mechanism (Mansard and Barthel (1984)).

In order to achieve a better understanding of the shoaling mechanism, and the relationship between offshore and nearshore wave climates within a Gaussian stochastic process, an extensive series of model tests was undertaken at the Hydraulics Laboratory of NRC. Waves were allowed to shoal on a slope of about 1:40 from an offshore depth of 26 m (full scale units) to a hypothetical breakwater site at 8.5 m depth. Using the random complex spectrum method of synthesis, fifty records of 20 min length, all derived from the same offshore spectral density, were simulated. Each of these records had somewhat different spectral characteristics because of the random complex spectrum method; their time domain characteristics in terms of wave statistics, groupiness, etc. were also different.

For each of the fifty records, the nearshore wave statistics were evaluated by computing fifteen different parameters describing the spectral and time domain features. In order to reduce the variability due to the relatively short record lengths of 20 minutes, longer term statistics were obtained by joining the data in blocks of 3 and 6 to give one and two hour averages respectively.

In order to evaluate the influence of the offshore wave heights on the statistical characteristics of nearshore wave climates, the above described simulations and analyses were repeated using four different offshore significant wave heights. Full details of this investigation can be found in Readshaw et al. (1987), and Mansard et al. (1988).

Figure 4 gives an example of the relationship between offshore and nearshore significant wave heights ($H_{1/3}$) obtained for the four different offshore sea states. This figure clearly illustrates that a large variability can be found in shoaled wave heights. The variabilty decreases as the record length increases, but even with 2 hour averages, there are substantial variations in the results. The figure also shows that a given nearshore wave height - for instance 6.5 m - can be obtained by a variety of offshore wave conditions, and is dependent on the chance selection of a particular time series.

The variations found in the offshore heights are partly due to differences in time domain characteristics of the synthesized waves. Wavebreaking in the offshore plateau and wave non-linearities also contribute to these variations.

Because the transformation of wave statistics during shoaling is not yet fully understood, and because of the serious implications caused by the chance selection of a particular offshore time series, a new approach is being developed by NRC for model studies which involve shoaling. This approach consists of determining experimentally the optimum (i.e. the worst) nearshore wave climate by testing a large number of time domain realizations for each offshore sea state. Figure 5 shows an example of the relationship between the offshore and the nearshore wave climates for ten different time domain realizations of a given offshore sea state. These time series, each with a record length of 45 minutes were synthesized, this time, using the random phase spectrum method in order to ensure the correct spectral shape. Since it is often unknown, as to which of the wave characteristics are critical in a model study, six different parameters considered relevant are generally computed. The choice of these parameters is based somewhat on intuition and could change depending upon the type of study. The particular offshore time series which gives consistently worst nearshore conditions for most of the six parameters is then selected as being the optimum wave condition to which the structure must be subjected. In this particular case the second sample appears to meet this criterion. Since the shoaling varies with the period and the height of the waves, a number of such selections may have to be done in a model study.

An alternative approach would be to test using very long records so as to ensure that all possible time domain variations of the offshore sea state occur. But design optimization of a given structure often requires a large number of test configurations, and therefore becomes expensive if very long records are used for each and every combination.

Evidently the best approach is to develop a better understanding of the shoaling mechanism and to evaluate the effect of parameters such as grouping, crest front steepness, variance of long waves, etc., pertaining to the offshore sea state, on the ultimate nearshore wave climate, so that a certain amount of prediction can be carried out without having to run all the tests. However, because of the complexity of the shoaling mechanism, a great deal of research remains to be carried out in this field.

5.0 CONCLUSIONS

- In the shallow water of the nearshore coastal zone, the presence of second order harmonics has to be taken into account when generating waves in model basins in order to obtain realistic response of coastal structures.

- The transformation of waves during shoaling is a very complex mechanism which requires that more extensive research be carried out.

- Although considerable progress has been achieved in the field of wave dynamics, propagation and transformation, a more thorough understanding of nature is still required before more realistic simulations in the laboratory are possible.

6.0 REFERENCES

1. Barthel, V., Mansard, E.P.D., Sand, S.E. and Vis.F.C., "Group Bounded Long Waves in Physical Models", Ocean Engineering Vol 10, No 4, 1983.

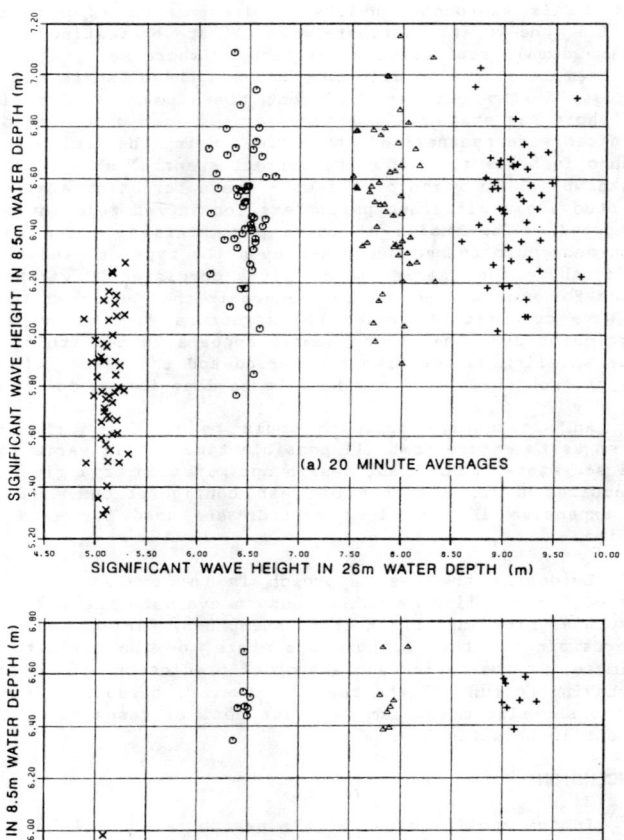

FIGURE 4

Transformation of significant wave heights (H13) due to shoaling
$T_p = 17$ s $\gamma = 3.3$

MODELING OF COASTAL STRUCTURES 19

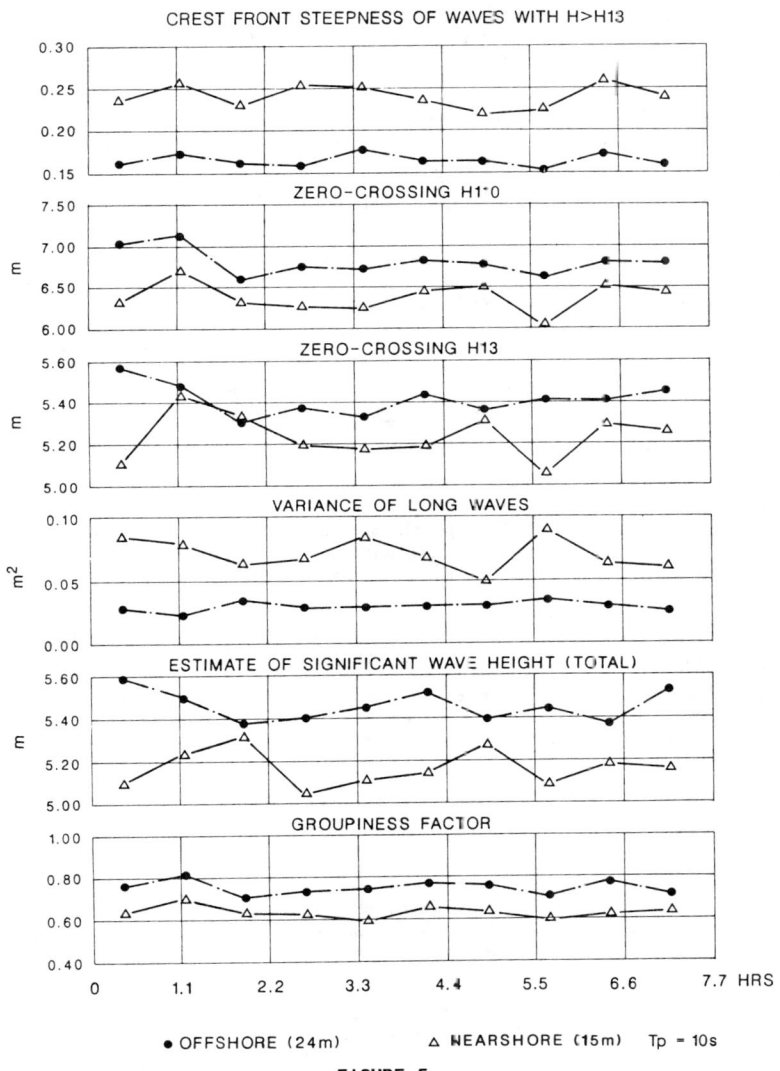

FIGURE 5

Relationship of selected wave parameters between offshore and nearshore
(Composite Slope 1:50 and 1:85)

2. Barthel, V., Mansard.,E.P.D., and Funke, E.R., "Effects of Group-Induced Long Waves on Wave Run-up", Proc. ASCE Coastal Structures, Arlington, VA, U.S.A., 1983.

3. Barthel, V. and Mansard E.P.D., "Second Order Waves -Importance in Experiment and Nature", Proc. IUTAM Symposium on Non-linear Water Waves, University of Tokyo, Japan, 1987.

4. Elgar, S., Guza, R.T. and Seymour, R.J., "Wave Group Statistics from Numerical Simulation of a Random Sea", Appl.Ocean Research, Vol. 7, No 2, 1985.

5. Funke, E.R. and Mansard, E.P.D., "On the Synthesis of Realistic Sea States", Hydraulics Laboratory Technical report, LTR-HY-66, National Research Council of Canada, 1979.

6. Funke, E.R. and Mansard, E.P.D., "The Control of Wave Asymmetries in Random Waves", Proc. 18th International Coastal Engineering Conference, Cape Town, South Africa, 1982.

7. Funke, E.R. and Mansard E.P.D., "The NRCC 'Random' Wave Generation package", Hydraulics Laboratory Technical Report TR-HY-002, National Research Council of Canada, 1984.

8. Funke, E.R. and Mansard, E.P.D., "A Rationale for the Use of the Deterministic Approach to Laboratory Wave Generation", Proc. Seminar on Wave Analysis and Generation in Laboratory Basins, 22nd IAHR Congress, Lausanne, Switzerland, 1987.

9. Funke, E.R. and Miles, M.D., "Multi-directional wave Generation with Corner Reflectors", Hydraulics Laboratory Technical Report, TR-HY-021, National Research Council of Canada, 1987.

10. Johnson, R.R., Mansard, E.P.D. and Ploeg, J., "Effects of Wave Grouping on Breakwater Stability", Proc. 16th International Coastal Engineering Conference, Hamburg, West Germany, 1978.

11. "List of Sea State Parameters", Joint Publication by the IAHR Section on Maritime Hydraulics and PIANC, Supplement to Bulletin No. 52 (1986), General Secretariat of PIANC, Residence Palace, ue de la loi 155, Brussels, Belgique.

12. Mansard, E.P.D. and Pratte, B.D., "Moored Ship Response in Irregular Waves", Proc. 18th International Coastal Engineering Conference, Cape Town, South Africa, 1982.

13. Mansard, E.P.D. and Funke, E.R., "A New Approach to Transient wave generation", Proc. 18th International Coastal Engineering Conference, Cape Town, South Africa, 1982.

14. Mansard, E.P.D. and Funke, E.R., "Variabilité statistique des paramètres de vagues", Proc. Int. Symp. on Maritime Structures in the Mediterrannean Sea, Athens, Greece, 1984.
(Also available in English as: "On the statistical variability of wave parameters", National Research Council of Canada, Hydraulics Laboratory Technical Report TR-HY-015, 1986.

15. Mansard, E.P.D. and Barthel, V., "Shoaling Properties of Bounded Long Waves", Proc. 19th Int. Conf. on Coastal Engineering, Houston, Texas, U.S.A. 1984.

16. Mansard, E.P.D., Sand, S.E. and Klinting, P., "Sub-and Superharmonics in Natural Waves", Proc. 6th Int. Symposium on OMAE, Houston, Texas, U.S.A., 1987.

17. Mansard et al, "On the Transformation of Wave Statistics Due to Shoaling", Hydraulics Laboratory Report in progress, 1988.

18. Medina, J.R. and Aguilar, J., "Comments on Numerical Simulation of a Random Sea: A Common Error and its Effect Upon Wave Group Statistics", Letters to the Editor, Applied Ocean Research, Vol. 7, No. 2, 1985.

19. Miles, M.D., Laurich, P.H. and Funke, E.R., "A Multi-mode Segmented Wave Generator for the NRC Hydraulics Laboratory", Proc. 21st, Am. Towing Tank Conf., Washington, D.C., U.S.A., 1986.

20. Miles, M.D. and Funke, E.R., "A Comparison of Methods in Synthesis of Directional Seas", Proc. 6th Int. Symp. on OMAE, Houston, Texas, U.S.A., 1987.

21. Myrhaug, D. and Kjeldsen, S.P., "Parametric Modelling of Joint Probability Density Distributions for Steepness and Asymmetry of deep water waves.", Appl. Ocean Research, Vol 6, No 4, 1984.

22. Pedersen, A.E., V. Barthel, C.S., Birt, "Port Development for St-Louis, Sénégal, Proc. 2nd COPEDEC, Beijing, China, 1987.

23. Pinkster, J.A., "Numerical Modelling of Directional seas", Proc. Symp. on Description and Modelling of Directional Seas, Tech. University of Denmark, Copenhagen, 1984.

24. Readshaw, J.S., Baird, W.F., and Mansard, E.P.D., "Shallow Water Wave Generation - An Engineering Perspective", Proc. Seminar on Wave Analysis and Generation in Laboratory Basins, 22nd IAHR Congress, Lausanne, Switzerland, 1987.

25. Sand, S.E., and Mansard, E.P.D., "Description and Reproduction of Higher Harmonic Waves" National Research Council of Canada, Hydraulics Laboratory, Technical Report TR-HY-012, 1986.

26. Tucker, M.J., Challenor, P.G., and Carter, D.J.T., "Numerical Simulation of a Random Sea: A Common Error and its Effect Upon Wave Group Statistics", Applied Ocean Research, Vol. 6, No 2, 1984.

**REEF BREAKWATER RESPONSE
TO WAVE ATTACK**

by

John P. Ahrens, Aff. M., ASCE

Abstract

A reef breakwater is a low-crested rubble-mound breakwater without the traditional multilayer cross section. This type of breakwater, in essence, is a homogeneous pile of stone with individual stone weights similar to those used in the armour and first underlayer of conventional breakwaters. Because of their high porosity, reef breakwaters are suprisingly stable to wave attack and, at the same time, can dissipate wave energy effectively.

Résumé

Un brise-lames de type récif est un brise-lames en enrochements à crête basse qui ne comporte pas, en coupe transversale, les couches successives des ouvrages classiques. Ce type de brise-lames est essentiellement un entassement homogène de pierres individuelles d'une masse analogue à celle des pierres utilisées pour la carapace et la première couche des brise-lames classiques. En raison de leur grande porosité, les brise-lames de type récif résistent avec une stabilité surprenante à l'assaut des vagues tout en dissipant de manière efficace l'énergie des vagues.

Reef Breakwater Response To Wave Attack

by John P. Ahrens*, Aff. M., ASCE

Abstract

A method of predicting stability of a class of low-crested rubble mounds referred to as reef breakwaters is presented. Findings are based on a study which included an extensive series of physical model tests conducted with irregular waves. Stability is measured in terms of reduction in crest height due to wave attack. Reduction in crest height has less scatter than damage measured by the number of stones displaced and provides a direct link between stability and primary performance characteristic of wave transmission. A stability model is developed which can accurately predict degradation of the reef from zero-damage to very severe levels and for a wide range of wave conditions. An example is presented which illustrates the ability of the model to predict damage, which in turn is used to estimate wave transmission and reflection characteristics and energy dissipated by the reef when it is in equilibrium with wave conditions.

Introduction and Background

A reef breakwater is a low-crested rubble-mound breakwater without the traditional multilayer cross section. This type of breakwater, in essence, is a homogeneous pile of stone with individual stone weights similar to those used in the armor and first underlayer of conventional breakwaters. Because of their high porosity, reef breakwaters are suprisingly stable to wave attack and, at the same time, can dissipate wave energy effectively.

It is anticipated reef breakwaters would be used primarily for beach stabilization or to protect eroding shorelines in a shore parallel or near parallel manner. Generally these functions can allow, or possibly benefit from, larger transmitted wave heights than can be tolerated in a harbor. This suggests that considerable cost savings could be achieved by using lower crested structures. In a discussion of a variety of typical coastal erosion problems Fulford (1985) concludes reef breakwaters represent the most satisfactory general approach to shoreline stabilization.

One of the recurring problems faced by coastal engineers has been to demonstrate that a stable rubble-mound configuration can be achieved with stone available. The fact that this could not always be done led to the development of concrete armor units. However, it was realized by some engineers that the full potential of stone was not being utilized. It has been frequently noted that the stability of a rubble-mound increases as it adjusts to wave attack (Bruun and Johannesson, 1976 and Bruun, 1985). Bruun has shown how some mature breakwaters have reached an equilibrium profile with relatively small

*Oceanographer, U.S. Army Engineer Waterways Experiment Station, P.O. Box 631, Vicksburg, Mississippi 39180-0631

stone, considering the wave climate. A logical extension of Bruun's work has been developed by Baird and colleagues through the use of an extensive, highly permeable berm incorporated into the breakwater cross section (Baird and Hall, 1984 and Baird and Hall, 1987). A highly porous berm helps distribute wave forces and dissipate energy. This strategy allows the use of smaller stone than would be required for armor of a traditional design which often means that construction can be conducted with land based equipment in a dump and push mode. Because of the tendency to develop a breakwater design with a traditional cross section and profile, concrete armor units have been used in situations where they were not required (Baird and Hall, 1984). This tendency may increase the cost of a rubble-mound and at the same time reduce it's reliability. Baird and colleagues have now refined the berm breakwater approach and have shown that substantial cost savings can be obtained without loss of functional performance or compromise on safety. Research by van der Meer and Pilarczyk (1987) has produced effective methods to parameterize the process where rubble mounds adjust to wave conditions. From this effort, a mathematical model was developed which can predict the equilibrium profile of a berm breakwater.

Since reef breakwaters would ordinarily be used to protect beaches and eroding shoreline, short periods of relatively high wave transmission could be accepted as part of a cost effective design. A logical approach, consistent with the findings of Bruun and Baird, to the design of reef breakwaters would be to allow or accept considerable adjustment of the profile so long as wave transmission characteristics fall within acceptable limits. Unfortunately, until recently the information necessary to adopt this strategy has not been available. A flexible approach to the design of reef breakwaters is presented in this paper which allows consideration of the response of the reef to wave action and it's resulting influence on the reef's performance. This approach is based on a synthesis of the findings of many other investigators and on an extensive series of laboratory model tests conducted at the Coastal Engineering Research Center (CERC). From these tests, a conceptual model for the stability of reef breakwaters evolved which finally led to the development of a mathematical model to predict the stability to wave attack. It is commonly observed that low-crested rubble mounds respond to severe wave action and heavy overtopping by having their crest heights reduced and their side slopes flattened (Carver and Davidson, 1983). This characteristic is a conspicuous response of reef breakwaters to wave attack and the response noticeably increases resistance to further damage. The reef breakwater stability model has the ability to predict this response accurately over a wide range of wave conditions and structure configurations.

Using the stability model to predict an equilibrium crest height for the reef allows this information to be used to estimate hydraulic performance characteristics of wave transmission and reflection and energy dissipation. Prediction of these characteristics are quite dependent on the stability model since hydraulic performance is so sensitive to crest height. The stability model will be discussed and its use for predicting hydraulic performance characteristics of reef breakwaters will be illustrated. Results from this research will help alleviate the problem noted by Dally and Pope (1986) about the lack of quantitative design guidance for detached breakwaters used to protect shorelines and beaches.

Laboratory Setup, Conditions, and Procedures

Reef breakwater model tests were conducted in a 61 cm wide channel within CERC's 1.2 m high by 4.6 m wide by 42.7 m long wave tank. Water depths at the reef were 25 cm for most tests and 30 cm for a few tests. Water depths at the wave generator were 25 cm greater than at the reef and waves shoaled over a 1 on 15 slope to reach the concrete platform the reef was built on. This setup insured that very severe wave conditions could be produced at the structure. The testing channel was open to a wave absorber area on the landward side of the reef so there could be very little ponding effect behind the breakwater to complicate evaluation of stability.

All tests were conducted with irregular waves. Spectra were JONSWAP type in the deeper portion of the wave tank prior to wave breaking. Incident and reflected spectra were resolved with three parallel wire resistance wave gages in front of the reef by using the method of Goda and Suzuki (1976). Period of peak energy density of the spectra, T_p, ranged from about 1.45 to 3.60 seconds and the range of incident zero-moment wave heights, H_{mo}, was about 1.1 to 18.2 cm. Two gages were used behind the reef to measure transmitted wave heights.

Two basic types of tests were conducted during the laboratory investigation. Most tests fell into a category where the primary objective was to determine stability of the reef to a specific wave condition and wave transmission data were obtained as a by product. These are referred to as stability tests. The other category was tests conducted at the completion of a stability test to determine transmission characteristics of a damaged structure to a variety of less severe wave conditions. These are called "previous damage" tests. Table 1 organizes the tests into subsets and gives important characteristics of each subset. In Table 1, stability tests have odd subset numbers and "previous damage" tests have even numbers. For stability tests, the stone was dumped in the dry testing channel and pushed around primarily by foot to conform to a template outlining the desired profile of the structure. This procedure was used to prevent overly careful placement of the stone, but at the same time to insure the initial profile of all reefs within a subset would be reasonably consistent. Desired initial profile for a stability test is a trapazoid with a crest width of three stone diameters and having both seaward and shoreward slopes of 1 on 1-1/2. When the profiles were surveyed, initial crest elevations of reefs within a subset could vary as much as ±1.5 cm from the desired height. Duration of wave action for stability tests was one and one-half hours for tests with T_p of 1.45 sec and up to three and one-half hours for tests of T_p of 3.60 sec. This duration gives a total of between 3500-4000 waves

Table 1
Basic Data For Each Subset

Subset No.	No. of Tests	Water Depth, d_s (cm)	Crest Height as Built, h_c' (cm)	Median Stone Weight, W_{50} (gr.)	Area of Breakwater Cross Section A_t (cm^2)
1	27	25	25	17	1170
2	3	25	NA	17	1170
3	29	25	30	17	1560
4	12	25	NA	17	1560
5	41	25	35	17	2190
6	11	25	NA	17	2190
7	38	25	32	71	1900
8	26	25	NA	71	1900
9	13	30	32	71	1900
10	5	30	NA	71	1900

NA denotes not applicable to "previous damage" test series.

based on T_p and the reefs appeared to be at equilibrium with wave conditions well before the completion of a test. At the completion of a stability test, there was a final survey to document the equilibrium profile of the reef.

After some stability tests, previous damage tests would be conducted. Previous damage tests always used wave conditions less severe than the preceding stability test so there was almost no change in the profile during this type of test. Previous damage tests were not used to develop the reef stability model. Figure 1 shows the initial and equilibrium reef profile for a series of tests with progressively higher wave heights.

Two sizes of stone were used during this study. Characteristics of stone and associated gradation are given in Table 2. Ahrens (1987) provides extensive information on laboratory setup, conditions, and procedures.

Discussion of Findings

During this study, it was found that the stability of the reef was strongly dependent on the wave period. The following stability parameter was identified as the best way to characterize severity of wave attack on a reef breakwater where N_s^* is referred to as the spectral stability number (Ahrens 1987),

$$N_s^* = \frac{\left(H_{mo}^2 L_p\right)^{1/3}}{\left(\frac{W_{50}}{W_r}\right)^{1/3} \left(\frac{W_r}{W_w} - 1\right)} \tag{1}$$

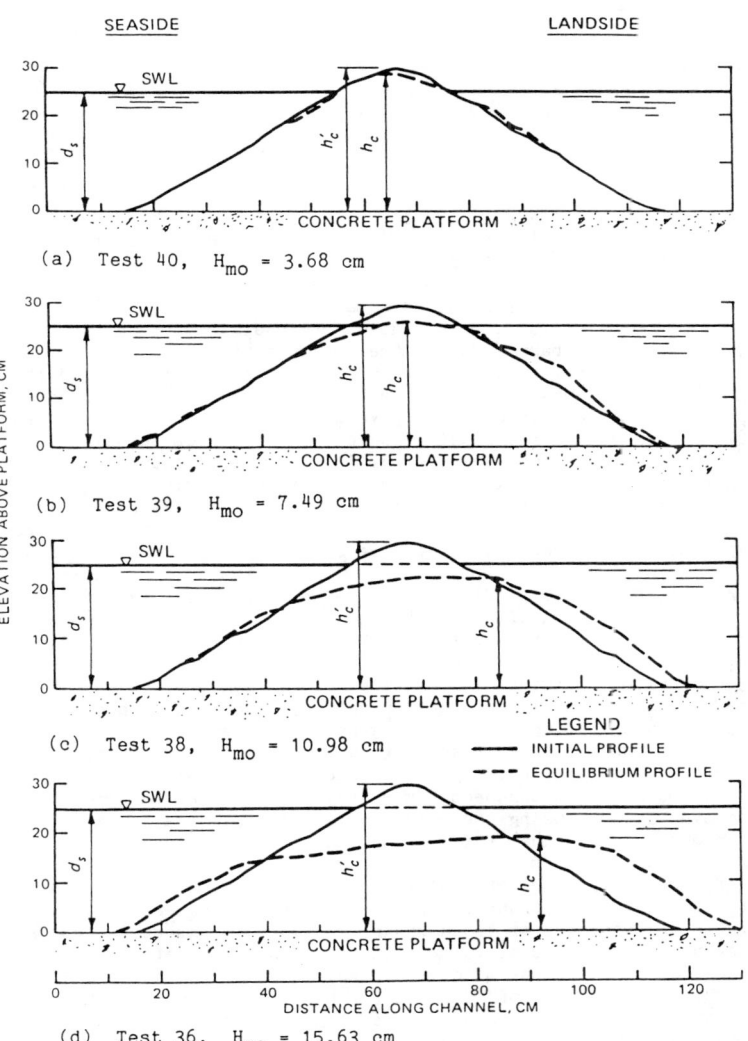

Figure 1. Reef profiles before and after wave action

Table 2

Stone and Gradation Characteristics

Characteristic	Quartzite	Diorite
2% weight (gr)	7.0	14.0
Median weight, W_{50} (gr)	17.0	71.0
98% weight (gr)	28.0	139.0
Density (gr/cm^3)	2.63	2.83
Porosity	44%	45%

where L_p is the Airy wave length calculated using d_s and T_p and W_w is the specific gravity of water. N_s^* is similar to the stability number developed by Hudson (1958) and used extensively to quantify the stability of traditional breakwaters (Hudson and Davidson, 1975). The need to account for the influence of wave period in the study of reef stability, as compared to earlier studies of traditional type rubble mounds, appears to be partly due to the use of a rather small wave height, H_{mo}, to characterize the stability, (i.e. small compared to the maximum wave height). Period effect also seems to be due to the fact that magnitude and intensity of overtopping flow are proportional to $(H_{mo}^2 L_p)$ as noted in the study of wave overtopping of seawalls, Ahrens, et al. (1986). Gravesen, et al (1980) gave an equation which contains the spectral stability number in their paper discussing irregular wave tests of breakwater stability. During this study, it was found that there was no net stone displacement for stability tests where $N_s^* \leq 6.0$ but there was noticeable stone movement for $N_s^* \geq 8.0$.

Figure 1 shows reef profiles for 4 tests. For each test, the profile at the beginning of the test (before any waves have been generated) is compared to the profile at the end of the test. The profile at the end of the test is regarded as the equilibrium profile. Table 3 summarizes important information about the tests shown in Figure 1. Tests shown in Figure 1 are intended to provide a typical sequence of how a reef responds to progressively higher waves. T_p is about the same for all four tests. One way to characterize response of the reef to wave attack is by use of reef response slope parameter, A_t/h_c^2. As wave conditions become more severe, the reef responds by becoming ever lower. Using the height of a reef to characterize stability is convenient since it is not so sensitive to random variations as using the number of stones removed or an equivalent method. The height of the reef also relates directly to its functional performance characteristics of wave transmission and reflection and energy dissipation. In the sequence of tests shown in Figure 1, the reef response slope A_t/h_c^2 has an initial value of about 1.80 for all tests. It increases from 1.92 for the mild wave conditions of Test 40 to 4.21 for the severe conditions of Test 36. A_t/h_c^2 is a measure of the seaward and shoreward slope of the reef (analogus to the contangent of a traditional breakwater) and can also be regarded as a shape variable.

RESPONSE TO WAVE ATTACK

Table 3
Conditions for Damage Sequence Profiles Shown in Figure 1

Subset/ Test No.	H_{mo}(cm)	T_p (sec)	A_t/h_c^2	N_S^*
3/40	3.68	2.79	1.92	5.91
3/39	7.49	2.82	2.33	9.53
3/38	10.98	2.81	3.13	12.28
3/36	15.63	2.98	4.21	15.86

In Figure 2, A_t/h_c^2 is plotted versus N_S^* for all 148 stability tests. A regression curve is fit to the data for $N_S^* \geq 6.0$; it is these tests that indicate an adjustment of the reef to wave conditions. The regression curve follows the trend of data for $N_S^* \geq 6.0$ surprisingly well. It is surprising because it is known that the relative height of the structure, h_c/d_s, and the bulk of the reef also play an important role in determining the amount of response of the reef to wave action (Ahrens, 1987). The equation of the curve shown in Figure 2 is given by

$$\frac{A_t}{h_c^2} = \exp [K\ N_S^*] \qquad (2)$$

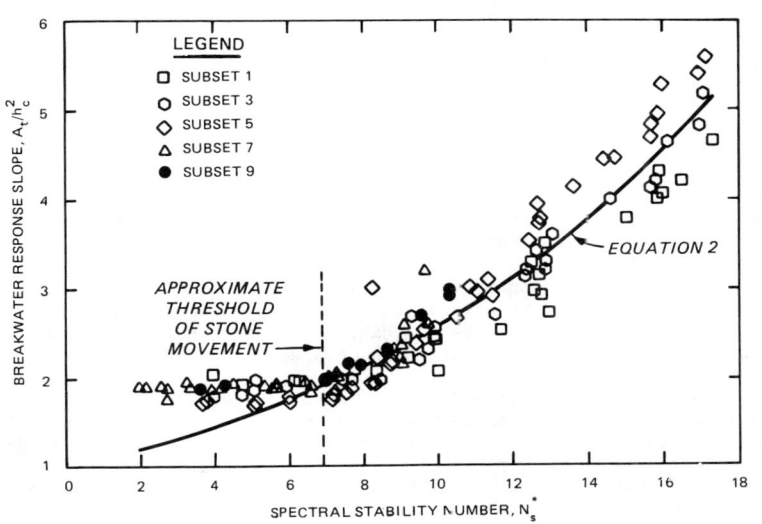

Figure 2. Reef breakwater response slope, A_t/h_c^2, versus the spectral stability number, N_S^*

where the regression coefficient is $K = 0.0945$ and is similar to a stability coefficient. Rewriting Equation 2 gives

$$\frac{1}{K} = \frac{N_s^*}{\ln\left(A_t/h_c^2\right)} \tag{3}$$

which resembles the form of Hudson's (1958) equation, i.e.

$$K_D = \frac{N_s^3}{\cot\theta} \tag{4}$$

where a stability coefficient is given as the ratio of a stability number to a function of the slope of the structure. Another advantage of the form of Equation 3, in addition to fitting the data well, is that it approaches logical limiting values

$$\frac{A_t}{h_c^2} \longrightarrow \infty, \text{ as } N_s^* \longrightarrow \infty$$

and

$$\frac{A_t}{h_c^2} \longrightarrow 1.0, \text{ as } N_s^* \longrightarrow 0$$

The latter limit is due to the fact that the angle of repose of stone is about 45 degree and therefore the smallest stable pile of stone in the absence of waves would be

$$\frac{A_t}{h_c^2} = 1.0$$

With this perspective, the regression curve in Figure 2 can be regarded roughly as a boundary between stable and unstable reef configurations. The data to the left of the curve for $N_s^* \leq 7.0$ are reefs which remained substantially in their original configuration due to the lack of sufficient wave action to displace stones, eg. the reef in Figure 1a. Figure 2 indicates that, for these conditions, the reefs would have been stable at somewhat steeper slopes.

When Figure 2 is examined, it can be seen that for $N_s^* > 7$ data from some subsets are consistently above the Equation 2 curve and data from other subsets are consistently below. For example, Subset 5 data are consistently above and Subset 1 data are consistently below the Equation 2 curve. To help explain these trends, Figure 3 shows the relative crest height of the reefs, h_c/d_s, versus N_s^*. In Figure 3, hand drawn curves are given for each subset consistent with the data showing progressive deterioration of the reef with increasing severity of wave attack. Figure 3 shows that, for any given value of N_s^*, the reefs of Subset 5 are higher than those of Subset 1. The

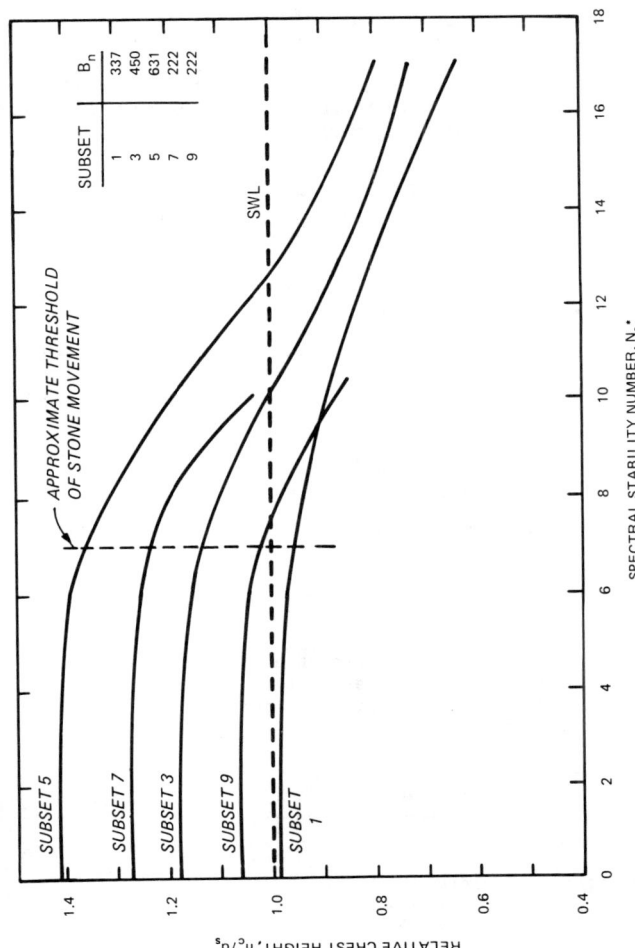

Figure 3. Damage trend curves showing the relative crest height, h_c/d_s, versus the spectral stability number N_s^*

greater stability of Subset 5 reefs is due to their flatter slope and greater bulk. Figure 2 indicates that the reefs of Subset 1 are stable at a steeper slope than those of Subset 5, for a given severity of wave action, due to their greater depth of submergence, as shown in Figure 3. The greater the depth of submergence, the more the reef is sheltered from wave forces. By using both Figures 2 and 3, the various influences on reef stability can be observed.

The implications of Figures 2 and 3, relating to reef stability or the reef's response to wave attack, can be summarized by generalizing Equation 2 in the following manner.

$$\frac{A_t}{h_c^2} = \exp\left[N_s^* \left\{ K\left(\frac{h_c'}{d_s}, \frac{h_c}{d_s}, B_n\right)\right\}\right] \quad (5)$$

$K\left(\frac{h_c'}{d_s}, \frac{h_c}{d_s}, B_n\right)$ indicates that the stability coefficient, K is regarded as a function of the relative crest height of the reef, both "as built" and at equilibrium to wave conditions, and the bulk or size of the structure as defined by the bulk number, B_n, where

$$B_n = \frac{A_t}{d_{50}^2} \quad (6)$$

The typical dimension of the stone in the reef is given by

$$d_{50} = \left(\frac{W_{50}}{W_r}\right)^{1/3} \quad (7)$$

Using regression analysis, a number of equations of the form of Equation 5 were obtained and investigated. The equation which fit the data best and reflected the physics, as currently understood, most satisfactory is given by

$$\frac{A_t}{h_c^2} = \exp\left[N_s^* \left\{ C_1\left(\frac{h_c' - h_c}{d_s}\right) + C_2\left(\frac{h_c}{d_s}\right)^{3/2} - C_3\left(\frac{h_c}{d_s}\right)^2 + \frac{C_4}{\sqrt{B_n}}\right\}\right] \quad (8)$$

where the regression coefficients are

$C_1 = 0.0460$

$C_2 = 0.2083$

$C_3 = 0.1440$

$C_4 = 0.4317$

Equation 8 is implicit in the equilibrium crest height and was found to converge rapidly using the as built crest height, h_c', as the initial value of h_c.

When the predicted equilibrium crest height, h_c, is compared to the observed crest height, it is found that Equation 8 provides very accurate estimates. Figure 4 shows the ratio of observed to predicted crest heights as a function of N_s^*. For $N_s^* > 7.0$ the predicted crest heights, with one exception, all are within ± 10 percent of the observed value. For $N_s^* < 7.0$ the reefs are not at the boundary of equilibrium with wave conditions and Equation 8 indicated that reefs somewhat higher or using somewhat steeper side slopes than the as built conditions would have been stable.

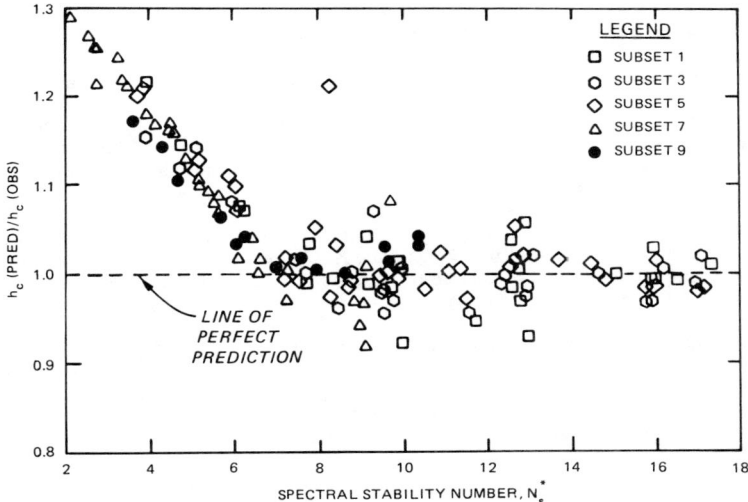

Figure 4. Ratio of the predicted crest height Equation 8 versus the observed crest height as a function of the spectral stability number

The purpose of the reef stability model is largely to be used to calculate wave transmission and reflection and energy dissipation characteristics by providing the equilibrium crest height of the structure. In Ahrens (1987), a simple method was developed to estimate wave transmission and reflection which was especially convenient for graphical presentation. This method selected the "best" formulation for wave transmission and reflection using two variables for both processes with one variable common to both. The functional relations selected were

$$K_t = f\left(\frac{h_c}{d_s}, B_n\right)$$

and

$$K_r = f\left(\frac{h_c}{d_s}, \frac{d_s}{L_p}\right)$$

where K_t and K_r are the transmission and reflection coefficients, respectively. Regression analysis was used to develop equations having the functional form noted above. The equations are

$$K_t = \frac{1.0}{1.0 + 0.0294 \left(\frac{h_c}{d_s}\right)^{3.3338} \left(B_n\right)^{0.5857}} \quad (9)$$

and

$$K_r = \exp\left[0.2899 \left(\frac{h_c}{d_s}\right) - 0.7628 \left(\frac{d_s}{h_c}\right) - 7.3125 \left(\frac{d_s}{L_p}\right)\right] \quad (10)$$

The predictive ability of Equations 9 and 10 is demonstrated in Figures 5 and 6, respectively. In Figures 5 and 6, the observed data are plotted on the ordinate and the predicted values on the abscissa.

Figure 7 demonstrates how the reef stability model, Equation 8, can be used with the transmission equation, Equation 9, to estimate the transmission coefficients. In Figure 7, it can be seen on the left how the height of the reef is reduced as the severity of the wave attack increases and the stability model follows this trend quite well. Using the estimated crest height from Equation 8 in the transmission equation, Equation 9, produces the predicted transmission trend curve on the right side of Figure 7. It can be seen that the predicted transmission trend follows the observed data very well.

Equations 9 and 10 can be used to construct a rough energy balance diagram for a reef. Using the reefs tested in Subsets 3 and 4 as the example, Figure 8 was constructed. For Figure 8, a value of $B_n = 450$ was used in Equation 9. In Equation 10, two wave reflection trends were generated using the largest and smallest relative depth tested during Subsets 3 and 4, i.e.

$$\frac{d_s}{L_p} = 0.045 \text{ and } 0.121$$

Figure 8 reflects the basic conservation of energy relation

$$K_t^2 + K_r^2 + \text{dissipation} = 1.0 \quad (11)$$

In Figure 8, the consequences of degradation of the reef can be seen. As the crest height is reduced by wave attack, transmission increases and reflection decreases. It can also be seen in Figure 8 that, as the reef is knocked down, the ability to dissipate wave energy is reduced. For a crest height of $h_c/d_s = 1.2$, the reef will dissipate about 80 percent of the incident short wave energy and about 60 percent of the longer wave energy. At a crest height of $h_c/d_s = 1.0$, the reef will dissipate about 70 percent of the incident short

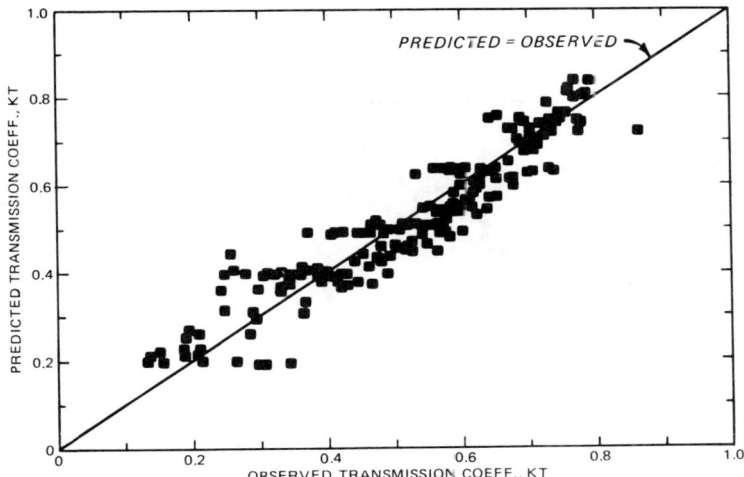

Figure 5. Predicted wave transmission coefficient Equation 9 versus the observed transmission coefficient

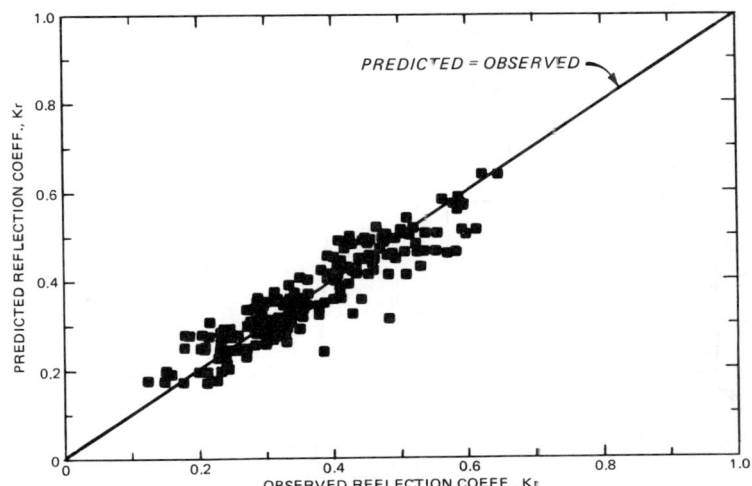

Figure 6. Predicted wave reflection coefficient Equation 10 versus the observed reflection coefficient

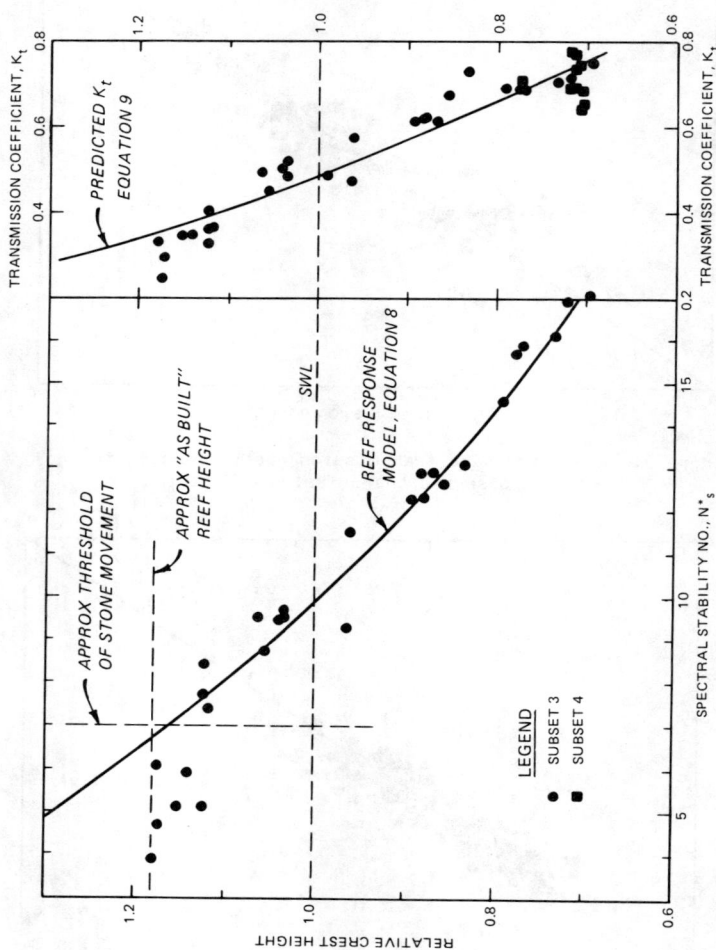

Figure 7. The reduction in height of the reef due to wave attack and the associated change in wave transmission

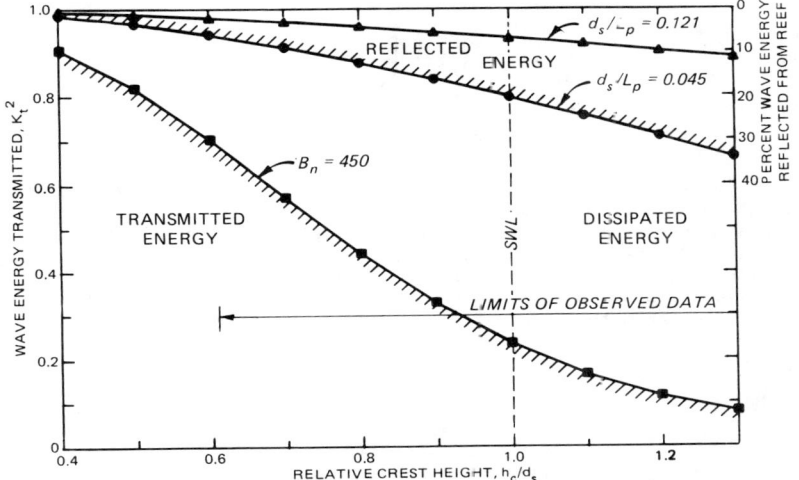

Figure 8. Wave energy balance due to the presence of a reef

wave energy and about 57 percent of the longer wave energy. In Ahrens (1987), methods are presented which allow better estimates of K_t and K_r than given by Equations 9 and 10 but they are more complicated and do not lend themselves to convenient graphical presentation such as Figure 8.

In using the reef stability model or figures such as Figure 8, a logical question would be how severe the wave conditions can be. It appears from wave tank tests that the limiting value for zero-moment wave height can be estimated as follows:

$$(H_{mo})_{max} \simeq (0.1) L_p \tanh\left(\frac{2\pi d_s}{L_p}\right) \quad (12)$$

(Ahrens and Heimbaugh, 1987). When Equation 12 is used in the equation for the spectral stability number, Equation 1, the following relation can be obtained

$$(N_s^*)_{max} \simeq \frac{\left(\dfrac{d_s}{d_{50}}\right)\left[0.1 \tanh\left(\dfrac{2\pi d_s}{L_p}\right)\right]^{2/3}}{\left(\dfrac{d_s}{L_p}\right)\left(\dfrac{W_r}{W_w} - 1\right)} \quad (13)$$

For the conditions of Subset 3

$$(N_s^*)_{max} = \frac{\left(\frac{25}{1.86}\right)\left[0.1 \times (0.2754)\right]^{2/3}}{(0.045)(1.63)}$$

$$= 16.7$$

Referring to Figure 7 it can be seen that the maximum observed value of N_s^* is about 17.0. Since the laboratory tests were conducted in such a manner as to encourage severe conditions, it appears that Equation 13 provides a reasonable estimate of the limiting conditions. Applying this information to Figure 8 indicates that the lowest relative crest height of the reef which could evolve due to natural wave action from the original trapazoidal shape would be

$$\frac{h_c}{d_s} \simeq 0.7$$

However, since it might not be convenient to build a reef with an initial configuration such as shown in Figure 1, it is interesting to extend the range of the relative height to values lower than could be obtained through natural evolution to evaluate the performance of an equivalent structure built, for example, by dropping stone from a barge.

Summary and Conclusions

It was found the stability of reef breakwaters was strongly dependent on the wave period such that longer period waves were more damaging than shorter period waves, other factors being equal. The most important variable influencing the stability of reefs was found to be the spectral stability number, N_s^*, defined by Equation 1. N_s^* is a good measure of the relative severity of wave attack on a reef. Other factors which affect the stability are relative crest heights of the reef, both as built and at equilibrium to the wave conditions, and the size of the structure as measured by the bulk number, B_n, defined by Equation 6. An analylitic expression, Equation 8, was developed for the stability of reef breakwaters which includes all variables mentioned above. Equation 8 fits the observed data very well and reflects the physics of the interaction between waves and reef as it is currently understood.

The stability model provided by Equation 8 can be used to predict the expected stable crest height of the reef for a wide range of wave and structure conditions. With this information, the wave transmission and reflection characteristics can be estimated using Equations 9 and 10. Knowing these characteristics, the amount of energy being dissipated by the reef can be calculated using the conservation of wave energy relation, Equation 11. Finally, a method for estimating the most severe conditions for a low-crested rubble-mound is given by Equation 13.

Acknowledgments

The tests described, unless otherwise noted, were conducted under the Coastal Structures Evaluation and Design program of the United States Army Corps of Engineers by the Coastal Engineering Research Center. Permission was granted by the Chief of Engineers to publish this information.

Appendix I. - References

Ahrens, J. P., "Characteristics of Reef Breakwaters," CERC Technical Report (in publication 1987).

Ahrens, J. P., Heimbaugh, M. S., and Davidson, D. D., "Irregular Wave Overtopping of Seawall/Revetment Configurations, Roughans Point, Massachusetts," CERC TR 86-7, Sep 1986.

Ahrens, J. P., "Reef Type Breakwaters," Proceedings of the 19th Coastal Engineering Conference, Houston, Texas, Sep 1984, pp 2648-2662.

Ahrens, J. P., and Heimbaugh, M. S., "Approximate Upper Limit of Irregular Wave Runup," CERC Technical Report (in publication 1987).

Baird, W. F. and Hall, K., "Breakwater Breakthrough," ASCE, Civil Engineering Magazine, Jan. 1987.

Baird, W. F., and Hall, K. R., "The Design of Breakwaters Using Quarried Stones," Proceedings 19th Coastal Engineering Conference, Houston, TX, Sep 1984.

Bruun, P., (editor), "Design and Construction of Mounds for Breakwaters and Coastal Protection," Elsevier, Amsterdam 1985.

Bruun, P., and Johannesson, P., "Parameters Affecting Stability of Rubble Mounds," Journal of the Waterways, Harbors, and Coastal Engineering Division, WW2, May 1976.

Carver, R. D. and Davidson, D. D., "Jetty Stability Study, Oregon Inlet, North Carolina," CERC TR 83-3, Sep 1983.

Dally, W. R., and Pope, J., "Detached Breakwaters for Shore Protection," CERC Technical Report 86-1, Vicksburg, MS, Jan 1986.

Fulford, E. T., "Reef Type Breakwaters for Shoreline Stabilization," Proceedings of Coastal Zone 85, Baltimore, MD., Sep 1985.

Goda, T., and Y. Suzuki, "Estimation of Incident and Reflected Waves in Random Wave Experiments," Proceedings of the 15th Coastal Engineering Conference, Honolulu, Hawaii, 1976, pp 828-845.

Gravesen, H., Jensen, O. J., and Sorensen, T., "Stability of Rubble-Mound Breakwaters," Presented at the Conference on Coastal Structures '79, Alexandria, VA 1979.

Hudson, R. Y., and Davidson, D.D., "Reliability of Rubble-Mound Breakwater Stability Models," Proceedings of the ASCE Symposium on Modeling Techniques, San Francisco, CA. 1975.

Hudson, R. Y., "Design of Quarry-Stone Cover Layers for Rubble-Mound Breakwaters; Hydraulic Laboratory Investigation," Research Report No. 2-2, U.S. Army Engineer Waterways Experiment Station, Vicksburg, MS, July 1958.

van der Meer, J. W., and Pilarczyk, K. W., "Dynamic Stability of Rock Slopes and Gravel Beaches," 20th Conference on Coastal Engineering, Tarpei, Taiwan, Nov. 1986.

Appendix II - Notation

A_t Cross sectional area of breakwater (cm^2)
d_s Water depth at toe of breakwater (cm)
d_{50} $(W_{50}/W_r)^{1/3}$, typical dimension of the median stone, (cm)
h_c' Crest height of breakwater as built (cm)
h_c Crest height of breakwater after wave attack (cm)
H_c Zero moment wave height at transmitted gage locations with no breakwater in channel (cm)
H_t Zero-moment transmitted wave height (cm)
H_{mo} Incident zero-moment wave height (cm)
K_r Reflection coefficient of breakwater as defined and calculated by method of Goda and Suzuki (1976)
L_p Airy wave length calculated using T_p and d_s (cm)
T_p Wave period of peak energy density of spectrum (sec)
W_r Density of stone, (gr/cm^3)
W_w Density of water, tests conducted in fresh water $w_w = 1.0$ (gr/cm^3)
W_{50} Median stone weight, subscript indicates percent of total weight of gradation contributed by stones of lesser weight (gr)
F   (H_c-d_s), freeboard of structure which for reef can be either positive or negative (cm)
K_t H_t/H_c, wave transmission coefficient
N_s^* Spectral stability number, defined by Equation 1
B_n Bulk number, defined by Equation 6
K   Stability coefficient for reef breakwaters
K_D Stability coefficient for conventional breakwaters
h_c/d_s Relative crest height

ROCK ARMOURING TO UNCONVENTIONAL BREAKWATERS: THE DESIGN IMPLICATIONS FOR ROCK DURABILITY

by

Allsop N.W.H. and Latham J.P.

Abstract

This paper identifies the need for quantitative durability testing of rock for use on dynamically stable rock armoured slopes. The derivation and use of a suite of engineering tests for rock quality is considered, and the limitations of simple accept/reject criteria discussed.

Recent developments in techniques for the quantification of armour slope profiles and armour unit shapes are presented. The use of a roller mill test to predict rock abrasion performance is discussed. The paper describes the use of advanced techniques to predict rock armour performance, allowing for long term reductions in armour unit size and roundness. The paper seeks to provide a framework within which to study the inter-relationships between wave climate, armour design, rock wear resistance, and weight loss and rounding in service.

Résumé

Cette étude reconnaît la nécessité d'essais quantitatifs concernant la durabilité des blocs de pierre à utiliser sur les pentes à carapace de blocs dynamiquement stables. On y prend en considération l'élaboration et l'utilisation d'un ensemble d'épreuves techniques pour déterminer la qualité de la roche et on y examine les limites de critères simples d'acceptation/rejet.

Des progrès récents des méthodes de quantification des profils de pente des carapaces et des formes des unités de carapace sont présentés. L'utilisation d'un essai au broyeur à rouleaux pour prédire la résistance de la roche à l'abrasion est examinée L'étude décrit l'utilisation de méthodes évoluées de prévision du rendement des carapaces de blocs de roches permettant des réductions à long terme des dimensions et de la rondeur des unités de carapace. Cette communication cherche à fournir un cadre pour l'étude des interrelations entre le régime des vagues, la conception des carapaces et la résistance à l'usure, la perte de masse ainsi que l'arrondissement des roches en cours d'exploitation.

Rock armouring to unconventional breakwaters: the design implications for rock durability

Allsop N W H+ & Latham J P*

+Principal Scientific Officer, Maritime Eng Dept Hydraulics Research, Wallingford, OX10 0BA, UK.

*Researcher, Centre for Applied Earth Science, Queen Mary College, Mile End Road, London, E1 4NS, UK.

Abstract

This paper identifies the need for quantitative durability testing of rock for use on dynamically stable rock armoured slopes. The derivation and use of a suite of engineering tests for rock quality is considered, and the limitations of simple accept/reject criteria discussed.

Recent developments in techniques for the quantification of armour slope profiles and armour unit shapes are presented. The use of a roller mill test to predict rock abrasion performance is discussed. The paper describes the use of advanced techniques to predict rock armour performance, allowing for long term reductions in armour unit size and roundness. The paper seeks to provide a framework within which to study the inter-relationships between wave climate, armour design, rock wear resistance, and weight loss and rounding in service.

1. Introduction

Much interest has recently been concentrated on 'berm', 'rock beach', and other types of breakwaters or sea walls on which small rock sizes are used to create a dynamically stable armour profile. The design of such structures allows wave action to move some of the armour up and down the slope until a profile is reached that is stable under the local wave conditions, but within which considerable armour movement may still continue.

Such a structure may be contrasted with the conventional rock armoured rubble slope, where the rock size and the slope angle at which it is laid together render the armour stable under the design conditions with minimal movement. Stochastic variations in the stability of individual armour units are permitted by allowing 'zero damage' to include up to 5% extractions from the slope. For such designs relatively large rock is required for static stability under all of the wave induced flows. The design condition for a conventional armour size may be given by a relative rock size, $H_s/\Delta D_n$ of around 1-2, but sometimes up to 4. In contrast the

relative rock size for a dynamically stable rock slope may be given by values of $H_s/\Delta D_n$ of between 3 and 20, where:
  H_s is the significant wave height of the design storm condition;
  Δ is the relative density $[(\gamma_r/\gamma_w)-1]$
  γ_r is the weight density of the rock;
  γ_w is the weight density of the (salt) water, often taken to be 1030 kg/m³;
  D_n is the nominal rock diameter defined in terms of the rock density and armour unit weight, W, $D_n = (W/\gamma_r)^{1/3}$

In general the median armour unit weight and nominal diameter, W_{50} and D_{n50}, will be used to characterise the armour.

This is a dramatic reduction in the armour size required, and has allowed a number of breakwaters to be built with locally derived rock armour which would otherwise have required the use of concrete armour units with the consequential increase in complexity, and perhaps cost. Examples of such structures have been discussed by Burren, Baird & Hall, Hall et al, and Gilman & Drage (Refs 1-4).

There is however one aspect of the performance of such structures that has received very little attention to date, that of the long term performance or durability of the rock armour. The increased movement of the armour, noted by Hall et al (Ref 3), may lead to significantly greater impact and abrasion damage to the armour than for the conventionally armoured structure with relatively immobile armour. In any case abrasion and/or impact will lead to both a reduction in the mean armour size, and to a change in armour unit roughness and shape. Each of these effects will in turn reduce the armour stability or, more correctly for a dynamically stable design, further increase the armour movement for a given wave climate or sea state.

Three alternative strategies may be postulated to deal with this problem.

(a) Rock of adequate resistance to abrasion and impact may be specified, and used if available. The derivation and use of engineering tests for rock durability have been discussed previously by Poole et al and Allsop et al (Refs 5,6). Example results of these tests have been presented by Allsop et al and Bradbury & Allsop (Refs 6,7).

(b) The locally available rock may be used, even if not in compliance with the pass/reject criteria implied in the acceptance values used in (a). The rates of rounding and weight loss may be estimated from on site measurements during the first few years life of the structure using techniques such as those discussed by Allsop et al, Bradbury & Allsop, and Fookes & Thomas (Ref 8). Predictions might then be made of the likely life of the armour, and of any maintenance or refurbishment that might be needed to extend the life of the structure. A simple example of this technique is discussed by Fookes & Thomas (Ref 8) who estimated the potential rate of deterioration of a new structure by quantifying the rates of rounding of armour on existing structures along the same coastline.

(c) The final approach would be for the designer to dimension the armour taking full account of possible levels of deterioration. Alternative designs could be prepared for each rock type or source available. The designer would then be able to explore the costs associated with the options of using locally available rock that might be of potentially low durability, in comparison with the alternative of importing more resistant rock at greater initial cost, but of potentially longer life and lower maintenance cost.

This strategy will require a knowledge of the overall wave climate, from which a measure of the lifetime energy level might be derived. The resistance of the rock type considered must be quantified, probably by a roller or tumbler mill test. It should then be possible to relate the armour movement anticipated in the design, the energy level of the wave climate, and the resistance of the armour, to the anticipated life of the armour.

The remaining sections of this paper will describe some of the recent developments, and give details of present work aimed at providing the data, and developing the techniques, needed by the designer of rock armoured coastal structures. Firstly we need to consider rock durability in the marine environment.

2. Rock quality and durability

Virtually all materials used for engineering purposes are required to conform to given limiting quantitative measures of strength and durability before they may be used. Engineering tests are routinely conducted on timber, steel, concrete, earth fill, and other materials to quantify their fitness for purpose, both in the short and long terms. Surprisingly rock armour used on sea walls and breakwaters is seldom specified to be other than 'hard and durable', or similar. It appears that only very infrequently are any significant engineering tests for quality and potential durability conducted on prospective rock sources.

This lack of quantitative measures of rock quality is particularly surprising in view of the often aggressive conditions to which rock armour on costal structures may be subjected. Impact and abrasion forces will tend to fracture and grind down the armour; salt crystallisation and freeze/thaw cycles will seek out small fractures and weaknesses; and a range of other mechanisms will combine to reduce the size, strength and resistance of the armour. A useful summary of the main mechanisms of degradation is given by Fookes and Poole (Ref 9). This subject is discussed in more detail by Allsop et al (Ref 6).

The extent of shape and roughness change may be illustrated by the armour shown in Figures 1a and 1b.

ROCK DURABILITY DESIGN IMPLICATIONS 45

Figures 1a Rock as placed in rip-rap armour

Figure 1b Abraded rock showing shape and roughness changes

Engineering tests for rock quality were first considered in a coherent fashion by Wakeling in a conference 10 years ago (Ref 10). Unfortunately this paper was not widely published. Since then one of the first steps in quantifying rock quality and relating it to durability was achieved by the research work carried out at Queen Mary College, University of London, and supported by Hydraulics Research, Wallingford. The results of that project have been discussed by Poole et al (Ref 5) and presented more fully by Allsop et al (Ref 6). The study derived data on rock deterioration from rubble breakwaters and sea walls in the United Kingdom, the United Arab Emirates, and Australia, spanning wide ranges of climatic and wave exposures. Preliminary results of roundness and weight loss measurements were discussed. A revised suite of engineering tests and appropriate acceptance values were tentatively suggested, together with a series of monitoring methods to quantify the state of a rubble slope in the field. These may be summarised:

Test	Recommended Acceptance Value
Aggregate impact value	25 maximum
Franklin point load index	$4MN/m^3$ minimum
Water absorption	2.5% maximum
Specific gravity	2.6% minimum
Magnesium sulphate soundness loss	12% maximum
Fracture toughness	$0.7MN/m^{3/2}$ minimum

Since then, work has continued to apply these techniques in practice, and to refine and develop methods for quantifying armour shape, the state of an armoured slope, and the resistance of rock to wear and rounding.

Some general example results for the engineering tests for 10 rock types were presented in Reference 6. The results of a number of site specific rock quality studies have been presented by Bradbury & Allsop (Ref 7), who considered 8 different potential rock sources. They also discuss the use of the test results in the design process, particularly alongside the results of hydraulic model tests. Bradbury & Allsop also report the use of the site monitoring techniques, first described in Reference 6, to quantify the state of the armoured slopes on three coastal structures. They estimate values of a total damage state by summing fractures, subsize armour, unstable armour and cavities; and they describe the likely changes to these values at various stages in the life of a structure.

At Queen Mary College considerable effort has been devoted to the development of new techniques to quantify armour slope profiles. These methods are intended to allow the assessment of the condition of the rock slope profile. Latham & Poole (Ref 11) have presented various methods of describing an armour slope profile using simple statistical functions of the profile height data, known as profile roughness descriptors. They describe the use of high and low pass filtering. The methods have been used most successfully with

laboratory profile data. A little field work has been conducted to
explore the use of these methods on full size structures. The
initial data rate demanded, at around 15-20 points per armour unit,
was impractical for manual surveying methods. A break of slope
method has been tried, reducing the data rate demanded by a factor
of 5, so that only around 4 points per unit were required. It has
not yet been possible to correlate any of the profile roughness
descriptors directly with hydraulic performance or armour layer
stability although objective measures of armour layer roughness
have been obtained. It is hoped that preliminary relationships
between hydraulic performance and roughness descriptions may be
derived using laboratory test data during autumn/winter 1987.

However, before any experiments can be conducted to explore the
effects of armour profile variations on hydraulic performance,
attempts must be made to identify those of armour shape. The first
step in this has already been taken in the development of automated
methods for particle shape measurement and analysis.

3. Armour unit shape, roundness and weight loss

The development of design methods to predict the extent and rate of
breakwater armour deterioration, and to quantify its effects on the
hydraulic performance, requires the quantitative description of
particle shape and texture. In developing new methods and
parameters, it is essential to ensure that:

(a) the method of measurement is repeatable and free from
subjectivity; and

(b) the parameters defined are meaningful, and capable of being
related to the performance of the rock in service.

In the initial stages of research on rock armour durability (Ref
6), the roundness of rock particles was described using methods
developed by Krumbein and Wadell. The roundness was given as the
average radius of curvature of the corners, expressed as a
percentage of the radius of the maximum inscribed circle. The same
method was used to quantify the roundness of rock armour units in
service, and aggregate particles used for laboratory tests of rock
quality and wear resistance. Changes of roundness in the field and
in a laboratory roller mill were then compared. This method is
however very subjective, and liable to variation.

More recently Latham and Poole have reported the development of a
revised roller mill test to quantify the wear resistance of rock
for use as armour (Ref 12). They have also developed methods to
quantify particle shape and texture using Fourier and Fractal
analysis techniques, and an automated video image processor to
digitise the particle outline (Refs 13,14). The revised roller
mill test is intended to quantify the wear resistance of potential
rock types and sources. The test is so designed that the results

should be indicative of the wear resistance of full size armour, although only aggregate size material is used.

The measurement of change of particle shape, and the weight loss during testing, together with the use of the test results, are discussed in more detail by Latham & Poole in Reference 14. Using Fourier analysis of the un-rolled particle outline they define a number of shape descriptor parameters. The Fourier Shape Factor and Fourier Asperity Roughness Factor, P_S and P_R, are defined over the first to tenth harmonics and the eleventh to twentieth harmonics respectively. Latham & Poole also discuss the use of the Fourier Total Roughness, Shape Contribution, and Surface Texture Contribution Factors. They suggest that weight loss can be related closely to a measure of particle shape, the Fourier Asperity Roughness factor, P_R. Values for this factor may be derived repeatably and simply using the video imaging system discussed in Reference 13. Example measurements of a range of shape and texture parameters are given for four samples of carboniferous limestone aggregate, originally drawn from a single stockpile. It is noted however that these descriptions of shape and texture have yet to be successfully applied to prototype armour. Techniques to measure shape parameters such as the Fourier Asperity Roughness from photographs of armour units in place on a breakwater or sea wall are presently under development. Illustrate particle outlines are shown in Figure 2.

4. Design of rock armoured slopes

The design of any armoured slope will require each of the main design parameters to be identified and its value calculated. At its simplest the size of armour needed, given by the median nominal rock diameter, D_{n50}, may be related to:

(a) the incident wave conditions;
(b) density of the rock available, and of the (salt) water;
(c) structure slope angle, and underlayer permeability;
(d) level of armour movement envisaged.

When taking armour deterioration into account the other parameter required is the wear resistance of the rock. The parameters listed in (a) and (d) should also indicate the severity of the forces acting to fracture and abrade the rock.

During the early field work reported by Poole et al (Ref 5) and Allsop et al (Ref 6), armour wear given by rounding was estimated for rock armour in environments of differing severity. Latham & Poole (Ref 14) define an equivalent wear factor, x, and using the field data suggest possible values. The equivalent wear factor may be defined as the number of years exposure at a particular zone on a given structure equivalent to one thousand revolutions in the roller mill. Using the estimated rounding values, Latham & Poole suggest a likely range for the wear factor of 0.5 to 5.0. Implicit in their definition of a factor for a given site is the assumption that this may in turn be related to the levels of incident wave energy and armour movement. The possible use of parameters based

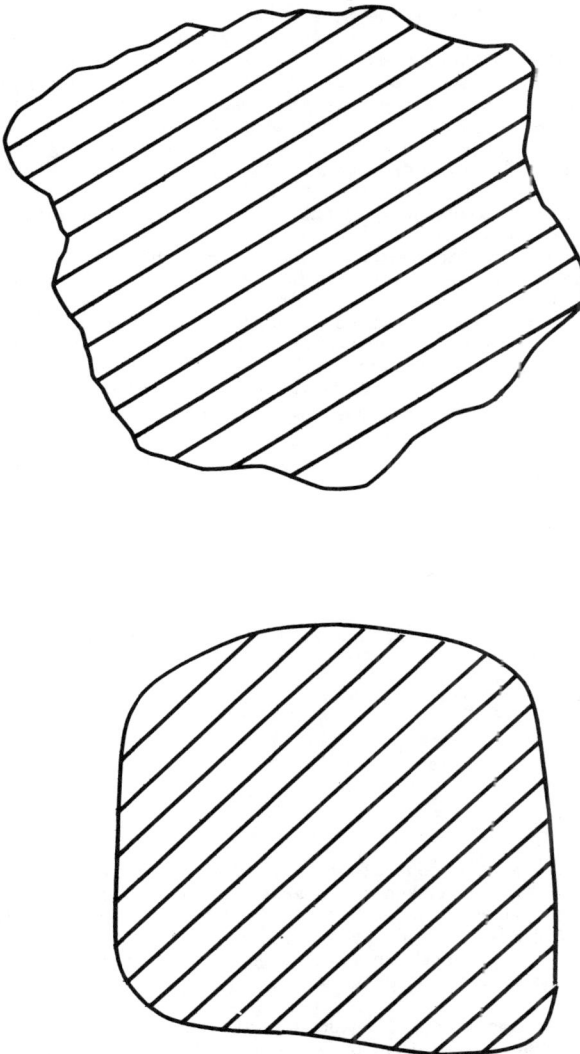

Figure 2 Outline of particles in Figure 1

on wave energy and comparative armour movement levels is discussed below. It should however be noted that none of the structures studied in the field work were designed for significant armour movement. It must be assumed that their design envisaged the usual minimal level of armour movement, and that armour rounding was primarily ascribable to the wave environment, rock type, and the other degradation mechanisms, not addressed in this paper. For dynamically stable slopes with larger relative wave heights $H_s/\Delta D_{n50}$, the greater level of armour movement will inevitably lead to increased wear. This might be well described by comparing the degree of armour movement with that on a statically stable slope.

The increased level of armour movement might then be described by calculating comparative values of the damage level, S, used by Van der Meer and Pilarczyk (Ref 17). This damage level may be determined from their equations relating armour size, D_{n50}, to the relative rock density, Δ; number of waves, N; notional underlayer permeability, P; incident wave conditions given by wave height, H_s, and Iribarren number, Ir; and damage level, S. Van der Meer & Pilarczyk derive two equations, the first for plunging wave conditions:

$$\frac{H_s}{\Delta D_{n50}} Ir^{\frac{1}{2}} = 6.2 \, P^{0.18} \, (S/\sqrt{N})^{0.2} \tag{1}$$

and for surging waves;

$$\frac{H_s}{\Delta D_{n50}} = \frac{Ir^P}{P^{0.13}} (\cot \alpha)^{\frac{1}{2}} (S/\sqrt{N})^{0.2} \tag{2}$$

These equations may be reversed to describe the damage level, S, under plunging waves:

$$S = \frac{N^{0.5}}{9161 \, P^{0.9}} \left(\frac{H_s \sqrt{Ir}}{\Delta D_{n50}}\right)^5 \tag{3}$$

and, under surging waves:

$$S = \frac{N^{0.5} \, P^{0.65}}{(\cot \alpha)^{2.5}} \left(\frac{H_s \, Ir^P}{\Delta D_{n50}}\right)^5 \tag{4}$$

For statically stable rock slopes the damage level adopted, $S = 2$, is equivalent to the Hudson 'no damage' criteria. A comparative armour movement parameter might therefore be defined by relating the level of damage S given by equations (3) and (4) as appropriate. This parameter might then be used to describe the contibution to increased armour wear.

The rate of armour wear will also depend upon the wave environment. This might be well described by a measure of the annual wave energy level, or of the design storm. This approach has already been considered by Behnke & Raichlen (Ref 15) and Timco (Ref 16), in relation to the breakage of concrete armour units, a not dissimilar problem! Timco calculates the energy level of the design storm, E_{inc}, as energy per wave per metre along the structure:-

$$E_{inc} = \frac{\rho_w g^2 H_s^2 T_p^2}{16\pi} \tag{5}$$

This definition implicitly assumes deep water at the structure, and normal wave attack. For a storm of duration, T_R, the total energy in the storm per metre front, E_{st}, may be defined:-

$$E_{st} = \frac{\rho_w g^2 H_s^2 T_p T_R}{16\pi} \tag{6}$$

Behnke & Raichlen (Ref 15) consider the energy of waves greater than a threshold value, H_{th}, for which no movement occurs. They propose a definition of a cumulative "energy", E_{cum}, which may be written:-

$$E_{cum} = \frac{C_{th} N L_p H_s^2}{2} \tag{7}$$

where the threshold coefficient, C_{th}, is defined:

$$C_{th} = [2 \left(\frac{H_{th}}{H_s}\right)^2 + 1] \exp\left[-2 \left(\frac{H_{th}}{H_s}\right)^2\right] \tag{8}$$

Assuming deep water, and a storm of duration, T_R, equation (7) may be re-written:-

$$E_{cum} = C_{th} \frac{g H_s^2 T_p T_R}{4\pi} \tag{9}$$

This expression is not however dimensionally consistent as a definition of energy, and it may be more useful to define energy for waves above the threshold, E_{th}, using equation (6) for energy in a design storm:-

$$E_{th} = C_{th} E_{st} \tag{10}$$

This then allows the rewriting of Behnke & Raichlen's expression in dimensionally consistent form:-

$$E_{th} = C_{th} \frac{\rho_w g^2 H_s^2 T_p T_R}{16\pi} \tag{11}$$

It may be noted that, when $H_{th} = 0$, $C_{th} = 1$, and

$$E_{th} = \frac{\rho_w g}{4} E_{cum} \tag{12}$$

The use of equation (11) to quantify the energy level of the design storm suffers a difficulty in setting an appropriate level for H_{th}. An approach to this is again suggested by the minimal damage level, $S = 2$, used in Van der Meer's equations, (1) and (2). Setting $S = 2$ and using the appropriate equations to calculate H_{th} will indicate the sea state above which significant armour movement may occur. This may then be used in calculating C_{th}.

In concluding it must be acknowledged that much of this arguement involves extrapolation and some speculation. No data on armour wear for structures with recently constructed dynamically stable armoured slopes is yet available. It is hoped that the damage level, S, and energy parameter, E_{th}, will provide indicators of levels of abrasive and fracture forces. These together with a measure of the wear resistance of the rock source considered, may contribute to a more satisfactory determination of the equivalent wear factor postulated by Latham & Poole (Ref 14). The mean yearly weight loss and roughness reduction of the armour blocks could then be predicted.

5. Conclusions and recommendations

Breakwaters or sea walls armoured with rock of relative size $H_s/\Delta D_n$ greater than about 3 may suffer from increased fracture and abrasion of the armour due to the higher level of armour movement. In turn these changes in armour size and shape will facilitate greater mobility of the armour.

A suite of engineering tests have been proposed, and have been used to test rock for potential durability. The field data from which these tests have been developed relates to statically stable slopes only. It is likely that dynamically stable slopes will require rock of higher specification if noticeable degradation is to be avoided.

The present suite of tests do not well test the abrasion resistance of the rock. A roller or tumbler mill test has been devised to simulate the main processes of abrasion and attrition. Advanced techniques have been developed to measure the shape and texture of abraded particles, and hence to quantify wear rapidly and repeatably. Good correlations have been found between changes in major shape parameters, such as Fourier Asperity Roughness, and particle weight loss. Methods are under development to allow shape and texture parameters to be measured directly from photographs of completed structures. It is hoped that this will permit the quantification of armour rock wear in service.

In addition to the wear resistance of the rock source used, the main factors influencing armour deterioration are the wave climate to which the structure is subjected; and the level of armour movement experienced on the structure. Simple empirical methods have been advanced for the quantification of each of these factors.

It is noted in the paper that some of the methods suggested are as yet not well supported by field data. The only field data available on armour rounding in the field was that measured by Queen Mary College in 1982. Designers and owners of rock armoured structures, particularly those of unconventional design, are strongly urged to measure the rates of weight loss and shape change with time. This field data will then allow the empirical framework for design suggested in this paper to be calibrated.

6. Acknowledgements

The work in the UK on which this paper is based has been funded in part by research grants from the Department of the Environment, the Ministry of Agriculture, Fisheries and Food, and the Science and Engineering Research Council. The authors are greatful for the advice and assistance of their colleagues, especially Dr A B Poole of Queen Mary College and Mr A P Bradbury of Hydraulics Research.

7. References

1. Burren K R. "Investigation, design and construction of a harbour on King Island, Tasmania". Proc Institution of Engineers, Australia, Civil Engineering Transactions, Paper 3458, 1975.

2. Baird W F, & Hall K R. "The design of breakwaters using quarried stones". Proc 19th Costal Eng Conf Houston, 1984.

3. Hall K R, Rauw C I, & Baird W F. "Wave protection for an offshore runway extension, Alaska". Proc Coastal Structures 83, Arlington, ASCE, March 1983.

4. Gilman J F & Drage B T. "St George harbour, a new design for Alaska's Bering Sea". Synopsis to 20th Coastal Eng Conf Taiwan, 1986.

5. Poole A B, Fookes P G, Dibb T E and Hughes D W. "Durability of rock in breakwaters". Proc Conf Breakwaters - design and construction, ICE, London, 1983.

6. Allsop N W H, Bradbury A P, Poole A E, Dibb T E & Hughes D W. "Rock durability in the marine environment". Report SR 11, Hydraulics Research, Wallingford, March 1985.

7. Bradbury A P & Allsop N W H. "Durability of rock on coastal structures". Proc 20th Coastal Eng Conf, Taipei, November 1986.

8. Fookes P G & Thomas R S. "Rapid site appraisal of potential breakwater rock at Qeshm, Iran". Proc ICE Pt 1, ICE, London, October 1986.

9. Fookes P G & Poole A B. "Some preliminary considerations on the selection and durability of rock and concrete materials for breakwaters, and coast protection works". Q J Eng Geol, Vol 14, 1981.

10. Wakeling H L. "The design of rubble breakwaters". Symp on design of rubble-mound breakwaters. Paper No 5, Experimental and electronic Laboratories. British Hovercraft Corp 1977.

11. Latham J P & Poole A B. "The quantification of breakwater armour profiles for design purposes". Coastal Engineering 10, Elsevier Science Publishers, Amsterdam, 1986.

12. Latham J P & Poole A B. "Pilot study of an aggregate abrasion test for breakwater armourstone". Q Jo Eng Geol London. 1987 (in press).

13. Latham J P & Poole A B. "The application of shape descriptor analysis to the study of aggregate wear". Q Jo Eng Geol London 1987 (in press).

14. Latham J P & Poole A B. "Testing the effects of abrasion on weight loss and surface roughness of rock fragments with application to breakwater armourstone degradation". Internal Report, Queen Mary College, London 1987.

15. Behnke D L & Raichlen F. "Breakwater armour displacement thresholds: a possible correlation with cumulative wave energy". Proc 19th Coastal Eng Conf Houston 1984.

16. Timco G W. "On the stability criterion for fracture in dolos armoured breakwaters". Coastal Engineering 8, Elsevier Science Publishers, Amsterdam, 1984.

17. Van der Meer J W & Pilarczyk K W. "Stability of breakwater armour layers: deterministic and probabalistic design". Delft Hydraulics Communication No 378, Emmeloord, February 1987.

8. Notation

C_{th} — wave energy threshold coefficient, defined in equation 8
D_n — nominal rock diameter defined as $(W/\gamma_r)^{1/3}$
E_{cum}, E_{inc}, E_{st}, E_{th} — wave energy levels defined in equations 5-1
g — gravitational acceleration, taken as 9.81 m/s²
H_{th} — threshold wave height
H_s — significant wave height
I_r — Iribarren number, defined as $\tan \alpha / (2\pi H_s/g T_m^2)^{\frac{1}{2}}$
L_p — wavelength of peak wave period, T_p, in deep water given as $g T_p^2/2\pi$
N — number of waves in a storm
P — notional permeability factor, see Reference 17
P_s — Fourier shape factor, see References 13,14
P_R — Fourier asperity roughness factor, see References 13,14
T_m — mean wave period
S — damage number, defined in Reference 17.
T_p — peak wave period
T_R — duration of (design) storm
x — equivalent wear factor, see Reference 14.
Δ — relative density $[(\gamma_r/\gamma_w)-1]$
α — slope angle of structure face
ρ_w — density of (salt) water
γ_w — weight density of water
γ_r — weight density of rock

ON THE STABILITY OF BERM BREAKWATER ROUNDHEADS AND TRUNK EROSION IN OBLIQUE WAVES

by

Hans F. Burcharth and Peter Frigaard

Abstract

The stability of a berm type breakwater (sacrificial breakwater) was tested in a 3-dimensional model at The Hydraulics Laboratory, Department of Civil Engineering, University of Aalborg.

The object was to study the stability/erosion of the breakwater head and the trunk, the latter exposed to both head-on and oblique irregular waves.

To avoid too many parameters a simple breakwater geometry and only one class of stones were used.

Résumé

La stabilité d'un brise-lames de type à risberme (brise-lames sacrificiel) a été éprouvée au moyen d'un modèle tridimensionnel au Laboratoire d'hydraulique du Département de génie civil de l'université d'Aalborg.

L'objet était d'étudier la stabilité/érosion du muscir et du tronc du brise-lames, ce dernier exposé à des vagues irrégulières frontales et obliques.

Afin d'éviter d'avoir à tenir compte d'un trop grand nombre de paramètres, le brise-lames était d'une géométrie simple et ne comportait qu'une seule classe de pierres.

RESHAPING BREAKWATERS.
ON THE STABILITY OF
ROUNDHEADS AND TRUNK EROSION IN OBLIQUE WAVES

by

Hans F. Burcharth[*]   Peter Frigaard[*]
Prof. of Marine Civil Engr.   M. Sc.

INTRODUCTION

The paper deals with the 3-dimensional stability of the type of rubble mound breakwaters where reshaping of the mound due to wave action is foreseen in the design. Such breakwaters are commonly named sacrificial types and berm types. The latter is due to the relatively large volume of armour stones placed in a seaward berm. However, as also conventional armoured breakwaters some times do contain a berm it is assumed that a better and more ambiguous designation would be "reshaping" rubble mound breakwaters.

The stability of a reshaping type breakwater was tested in a 3-dimensional model at The Hydraulics Laboratory, Department of Civil Engineering, University of Aalborg.

The object was to study the stability/erosion of the breakwater head and the trunk, the latter exposed to both head-on and oblique irregular waves.

To avoid too many parameters a simple breakwater geometry and only one class of stones were used.

MODEL TEST SET-UP

The stone material

The model consisted of one grading of crushed stones with a density ρ_s = 2.65 t/m³ (metric ton) and a gradation as given in Figs. 1 and 2.

It was found that the relationship between the sieve diameter (quadratic sieve) d and the stone volume V and stone weight w is

$$V = 0.7 d^3 = \frac{w}{\rho_s}$$

d is regarded a characteristic diameter of the stones.

---

[*] University of Aalborg, Department of Civil Engineering, Sohngaardsholmsvej 57, DK-9000 Aalborg, Denmark.

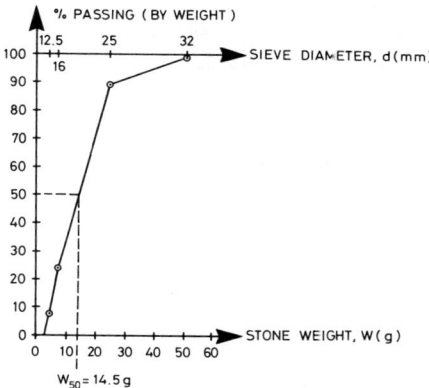

Fig. 1. Gradation of model stones (linear representation).

As seen from Fig. 1 w_{50} is found to be 14.5 g. However, it is most likely that an extra point on the gradation curve in the sieve interval 16-25 mm would have shifted the graph to the left and thereby given a w_{50} smaller than 14.5 g.

This is investigated by the log representation, Fig. 2.

The stone weight corresponding to $d_{50}^{\log} = 19$ mm is $w_{d_{50}} = \rho_s \cdot 0.7 d^3 = 2.65 \cdot 0.7 \cdot 1.9^3 = 12.7$ g $\geqslant w_{50}^{\log} = 12$ g.

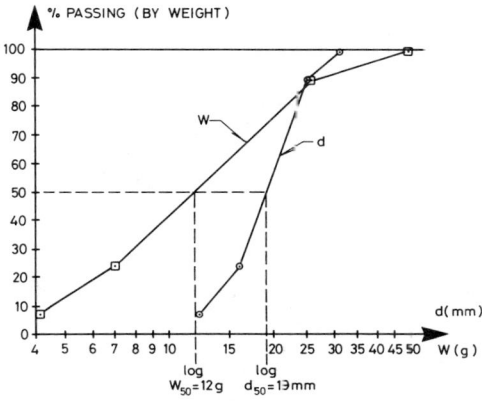

Fig. 2. Gradation of model stones (log - linear representation).

The log representations confirm that w_{50} is somewhat less than 14.5 g.

Based on the various figures a w_{50} of 12.7 g corresponding to d_{50}^{log} = 19 mm is chosen as the most correct value.

It should be noted that for the investigation of longshore transport in oblique waves samples of stones without diameters less than 16 mm were used. For these samples d_{50}^{log} = 22 mm and $w_{d_{50}}$ = 20.3 g.

The geometry of the model

Fig. 3 shows the cross section of the model (before each test) and the lay-out of the model in the wave basin.

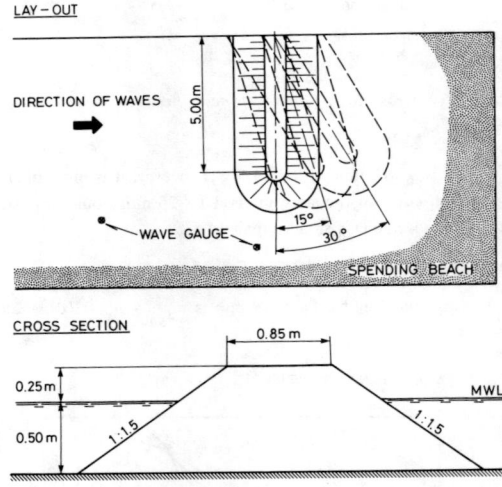

Fig. 3.   Lay-out and cross section of the model.

Waves

All waves were irregular waves generated in accordance with a random phase JONSWAP-type spectrum (peakedness parameter γ = 3.3 and width parameters σ_f = 0.10 for $f \leq f_p$ and σ_f = 0.50 for $f > f_p$).

The tested sea states are specified in Fig. 4.

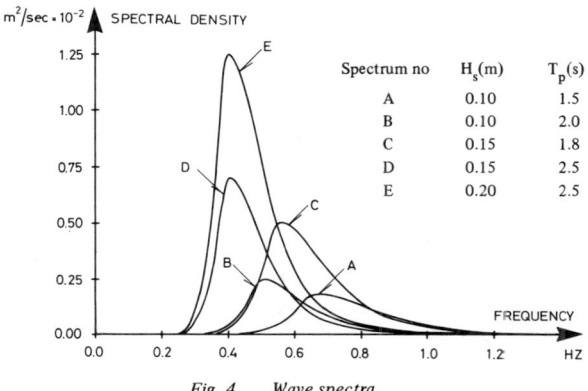

Fig. 4. Wave spectra.

The following three angles of attack were tested: $\alpha = 0$ (head-on waves), $\alpha = 15°$, and $\alpha = 30°$.

MODEL CONSIDERATIONS

The sea states were chosen in the range from very little erosion to fast erosion of the profile under oblique wave attack.

An indication of the relative stability of the profiles can be given by the dimensionless parameter $H_s/\Delta \cdot D_{n50}$, where $\Delta = \rho_s/\rho - 1$ and D_{n50} is a nominal diameter defined as $(w_{50}/\rho_s)^{1/3}$. It is seen that $H_s/\Delta D_{n50}$ equals the stability number $N_s = (K_D \cot\alpha)^{1/3}$, where K_D is the Hudson stability coefficient and α is the slope angle. (Note that the influence of the wave period os lacking in the parameter.)

According to extensive testing of rock slopes in head-on waves by DHI (Pilarczyk and Van der Meer) the values of the dimensionless parameter can be related to various types of rock slopes as follows:

$H_s/\Delta D_{n50}$

1 – 3	Conventional breakwater layer, start of damage
2 – 5	Conventional breakwater layer, failure
3 – 7	Berm breakwaters
5 – 50	Rock beaches

In the present tests we have

$$\Delta = (\frac{\rho_s}{\rho} - 1) = 1.65$$

$$D_{n50} = (\frac{12.7}{2.65})^{1/3} = 1.69 \text{ cm}$$

and consequently the range of tests corresponds to

$$3.5 < H_s/\Delta D_{n50} < 7.1$$

which is the interval considered for berm type breakwaters.

SCALE EFFECTS CONSIDERATIONS

Provided that the grading of the stones is not too wide, say $\frac{d_{85}}{d_{15}} < 3$, and provided that the amount of fine material cannot block the pores (which is usually the case if $\frac{d_{85}}{d_{15}} < 3$) it is relevant to define a Reynolds' number with the characteristic length D_{n50}

$$Re = \frac{D_{n50} \sqrt{gH_s}}{\nu}$$

ν is the kinematic viscosity = 10^{-6} m²/s at 20°C.

With H_s = 0.10 - 0.20 m we get

$$1.7 \cdot 10^4 < Re < 2.4 \cdot 10^4$$

Juul Jensen and Klinting analysed the scale effects and found that no significant viscous scale effect is to be expected if in the outer part of the structure $Re > 0.6 \cdot 10^4$. Van der Meer found no scale effects for rock slopes with characteristic stone size of 20 mm, which is approximately the stone size in the present tests. This is also the experience of the Hydraulics Laboratory at the University of Aalborg.

However, although it is believed that a viscous scale effect is present it will be either negligible or will cause the results (in terms of amount of damage) to be on the safe side.

TEST PROCEDURE

Stability of round head and trunk in head-on waves
The initial profile in each test was the one shown in Fig. 3.

The waves were recorded continuously throughout all the tests.

The breakwater profile was measured after N = 3000 waves in all tests and also after 6000 and 9000 waves in some tests.

Moreover, the characteristics of the stone movements were found from video recordings of the movements of coloured stones.

Stability of trunk and longshore transport in oblique waves

In a trial test series it was found that for a given H_s, T_p the dynamically stable profiles in oblique waves within the tested range $\alpha \leqslant 30°$, cf. Fig. 3, were almost identical to the profiles in head-on waves.

Thus in every test with oblique waves the initial profile was chosen as the one found after 3000 head-on waves.

Fig. 5. *Example of the breakwater before and after longshore transport tests in oblique waves.*

The longshore transport was found from video recordings of the movements of coloured stones placed in three bands over the profile. Moreover, after a specific number of waves (or time) the number, the positions and the total weight of each type of coloured stones were recorded.

The band width and the number of waves N were adjusted to the sea state in such a way that within the test period the non-coloured stones upstream the coloured bands did not pass the downstream coloured band. In this way the average transport per second (or per wave) through a cross section could be found. Moreover, by studying the distribution of the coloured stones over the profiles the maximum erosion depth (i.e. the number of stone layers within which displacements take place) could be estimated.

Fig. 5 shows photos before and after a test.

TEST RESULTS

Profiles in head-on waves

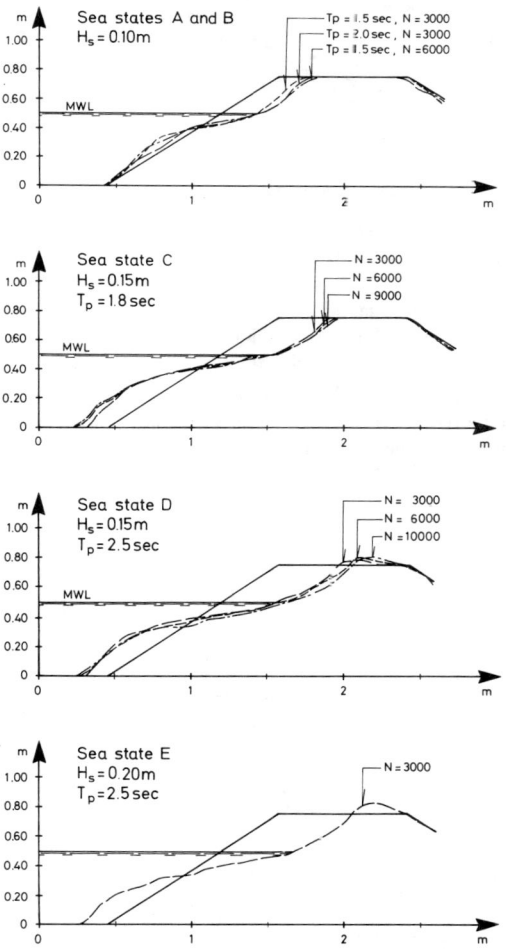

Fig. 6. *Profiles in head-on waves.*

Fig. 7 shows the various profiles for N = 3000.

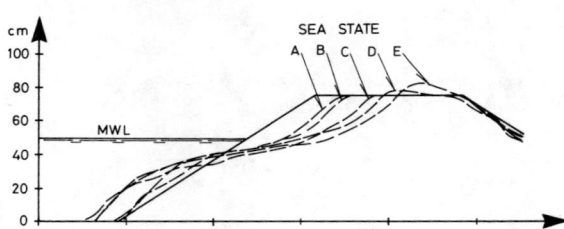

Fig. 7. Comparison of profiles after 3000 waves.

The material deficit is due to settlements caused by wave compaction and material transport across the crest.

Stability of the breakwater roundhead

The erosion of the roundhead is expressed in terms of the rate of recession of the crest measured along a longitudinal line parallel to the centerline of the breakwater, see Fig. 9.

Fig. 8 shows the recession as function of the number of waves. It is seen that the rate is almost constant for a certain sea state, i.e. a linear relationship between the recession of the crest end and the time (or number of waves).

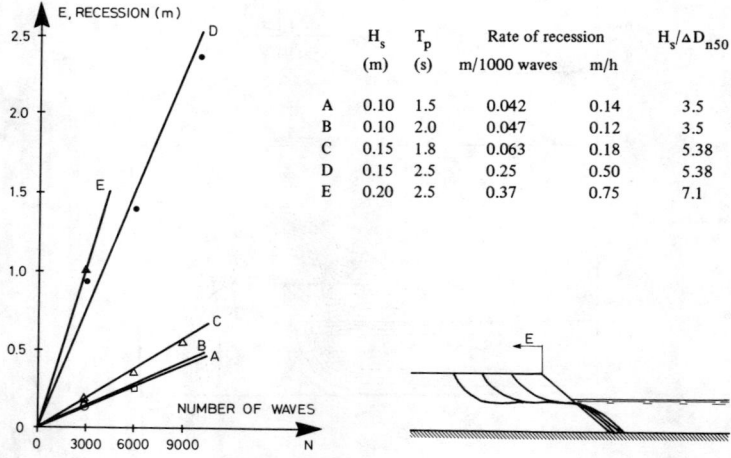

	H_s (m)	T_p (s)	Rate of recession m/1000 waves	m/h	$H_s/\Delta D_{n50}$
A	0.10	1.5	0.042	0.14	3.5
B	0.10	2.0	0.047	0.12	3.5
C	0.15	1.8	0.063	0.18	5.38
D	0.15	2.5	0.25	0.50	5.38
E	0.20	2.5	0.37	0.75	7.1

Fig. 8. Recession (erosion) rate of the roundhead.

It is seen from Fig. 8 that the roundhead erosion rate is small up to a certain sea state. When this is exceeded erosion is fast and the sea state seems to be characterized by the ability of practically every one of the waves to erode some of the stones from the roundhead and to displace (shift) them all the way across the roundhead.

Fig. 9 shows examples of the eroded longitudinal profiles as well as the bathymetry of the roundhead. A characteristic banana shape more pronounced than shown in the figure develops as the erosion proceeds.

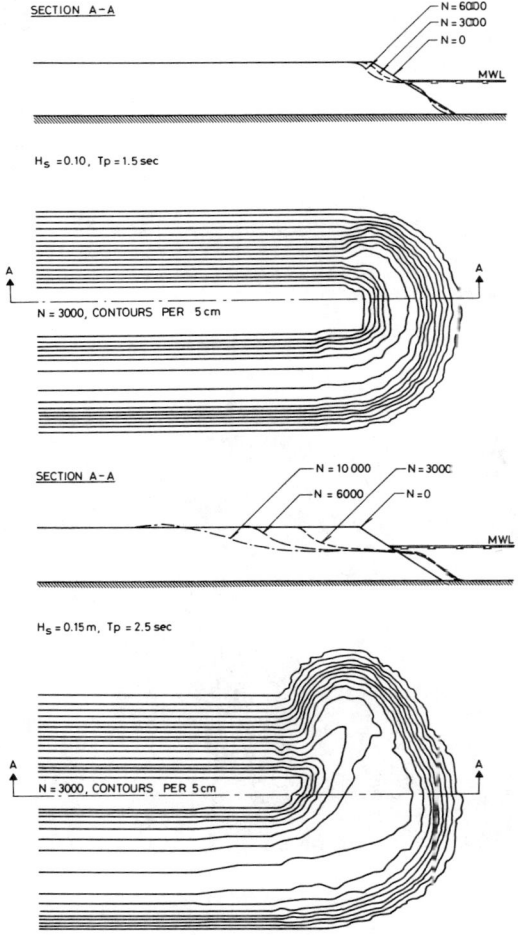

Fig. 9. Examples of erosion of roundhead.

Fig. 10. Photo of roundhead after exposure to 3000 waves characterized by $H_s = 0.10$ m and $T_p = 2.0$ s.

Fig. 11. Photo of the roundhead after exposure to 3000 waves characterized by $H_s = 0.15$ m and $T_p = 2.5$ s.

Fig. 8 shows the measured recession, E, of the model as function of the number of waves. In order to make a simple non-dimensional representation for the recession a "Shields approach" is used for the examination of the governing parameters.

The sediment transport close to the bed is usually described in a dimensionless form by the Meyer-Peter formula

$$8(\theta - \theta_c)^{3/2} = \frac{q_b}{\sqrt{\Delta \cdot g \cdot d_3}}$$

where

$\theta = \dfrac{\tau}{\rho \cdot \Delta \cdot g \cdot d}$ is the dimensionless bed shear

q_b = transport

$\Delta = \rho_s/\rho - 1$, ρ_s = density of stone

ρ = density of water

d = average grain size

g = gravity

θ_c = critical dimensionless bed shear stress

τ = shear stress

The bed shear stress τ can be expressed as a function of the flow velocity amplitude near the bed.

$\tau = \dfrac{1}{2} f_\omega \, \rho \cdot u^2$

f_ω = Jonsons wave friction factor

u = amplitude of flow velocity

Considering the shallow water in the breaker zone on the slope the flow velocity is assumed to be proportional to $\sqrt{g \cdot H_s}$.

In the Meyer-Peter formula the bed transport is related to $\sqrt{\Delta \cdot g \cdot d^3}$ in order to get a dimensionless expression for the bed transport. In the present experiments a formula for the erosion of the roundhead in terms of the recession E (of Fig. 8) is saught. Dividing the recession E with the deep water wavelength $\ell \cong g \cdot T^2$ a dimensionless recession is found.

Applying in principle the Meyer-Peter formula we obtain

$$\frac{E}{\ell} \cong (\theta - \theta_c)^{3/2}$$

As a first simplified approach the bed transport (erosion) is related to the characteristic sea state parameters H_s and T_z

$$\frac{E}{g \cdot T_z^2} \cong \left(\frac{\frac{1}{2} f_\omega \, \rho \cdot g \cdot H_s}{\rho \cdot \Delta \cdot g \cdot D_{n50}} - \theta_c \right)^{3/2}$$

$$E \cong \frac{T_z^2 \cdot g \, (H_s - \Delta \cdot D_{n50} \cdot \theta')^{3/2}}{\Delta^{3/2} \cdot D_{n50}^{3/2}}$$

The last expression is based on the simplifying assumption that Jonson's wave friction factor f_ω can be taken as constant. f_ω may be calculated using for example Peter Nielsens expressions or Bijkers expressions. However, because the flow over the bed is composed of oscillatory flow and current, and because the flow is highly influenced by the wave breaking it is rather complicated to determine a meaningful f_ω.

f_ω is a function of a/k, i.e. the water particle displacement divided by the bed roughness, but the variation is weak. In a normal breakwater situation a/k will probably show small variations and therefore the variation of f_ω will be marginal.

By applying a least square fit the threshold value θ' is found to be 0.82. This value may look small compared to a threshold value for stone movements estimated from the Hudson formula $H_s = (K_D \cdot \cot\alpha)^{1/3} \cdot \Delta \cdot D_{n50}$. However, it has to be remembered that the threshold value is related to a characteristic wave height, H_s, which is an over-simplification as it does not reflect correctly the effect of an irregular wave train. Fig. 12 shows the dimensionless recession.

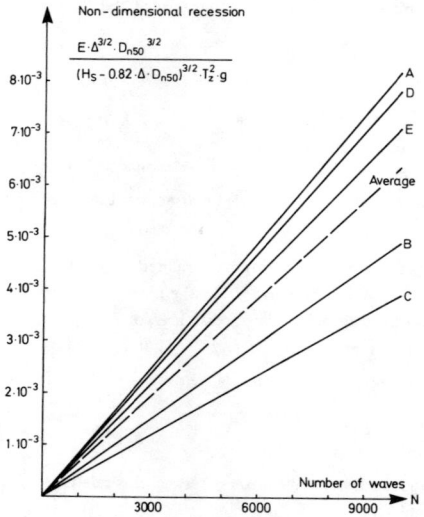

Fig. 12. Non-dimensional recession (erosion) of the roundhead).

It is seen that a completely satisfactory universal representation is not obtained. However, until more tests and further analyses are completed Fig. 12 might provide a crude first estimate of roundhead recessions.

In order to utilize the test results for a crude estimation of roundhead erosion also for breakwaters with a geometry somewhat different from the one presented here it is suggested to consider the resistance per unit volume stone independent on the cross sectional geometry (of course within some limits).

Having observed in the tests that the erosion reached a level of approximately $H_s/2$ below the still water level the dimensionless cross section area $A^* = A/(\Delta \cdot D_{n50})^2$ is introduced, where A is defined in Fig. 13.

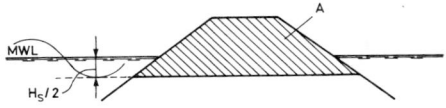

Fig. 13. Definition of eroded cross sectional area, A.

The prototype recession E_p of some structure might then be expressed in terms of:

E_{AU} — the recession found using Fig. 12 (corresponding to a geometry similar to the Aalborg University (AU) model)

A^*_{AU} — dimensionless cross sectional area of test geometry

A^*_p — dimensionless cross sectional area of prototype

$$E_p = \frac{A^*_{AU}}{A^*_p} \cdot E_{AU}$$

$$\cong \frac{450 \cdot E_{AU}}{A^*_p}$$

Based on the presented model tests and the behaviour of some prototype breakwaters the following somewhat premature recommendations valid for permanent designs are proposed:

	$H_s/\Delta D_{n50}$
For trunks exposed to steep oblique waves	< 4.5
For trunks exposed to long oblique waves	< 3.5
For roundheads	< 3

These values should be used as guidelines *only* if no other more quilified information is available. This is because the parameter $H_s/\Delta D_{n50}$ is insufficient as it among other things do not contain the effect of wave length and the effect of the duration of the sea.

It is believed that in additional investigations of erosion of reshaping breakwaters it will be necessary in principle to examine and summarize the respons from every single wave instead of using characteristic parameters like H_s to characterize the sea state. This is because the character of the flow kinematics in the erosion zones are strongly dependend on the size and the steepness of the single waves.

Effect of oblique waves on the trunk

The steady state transport of stones along the trunk was studied for two angles of wave attack, $\alpha = 15°$ and $30°$, cf. Fig. 3.

The results are given in Table 1.

Table 1. Steady state mass transport along the trunk.

Test no	Sea state	H_s cm	T_p sec.	α deg.	Average mass transport, Q	
					g/wave	g/sec.
6	A	10	1.5	15	0.48	0.45
7	B	10	2.0	–	0.95	0.68
8	C	15	1.8	–	14.7	11.7
9	D	15	2.5	–	12.4	7.1
10	E	20	2.5	–	56.0	32.0
11	A	10	1.5	30	0.78	0.74
12	B	10	2.0	–	1.29	0.92
13	C	15	1.8	–	20.5	16.3
14	D	15	2.5	–	30.7	17.5
15	E	20	2.5	–	62.5	35.7

It is seen that the sea states A and B only cause very small mass transports corresponding more or less to the onset of long-structure transport. It is also seen that the more severe sea states result in a significant transport which in typical prototype situations will change the geometry of the structure and might endanger the stability.

The results look homogeneous in the sense that the transport increases with wave height, wave period and angle of incidence in the tested ranges. The exception is test no 9 where the transport seems to be unexpected low.

The following formula for longshore transport on gravel beaches was given by van Hijum et al. 1982:

$$\frac{S}{gD_{90}^2 T_s} = 7.12 \cdot 10^{-4} \frac{H_{sd} \cos^{1/2} \alpha}{D_{90}} \left(\frac{H_{sd} \cos^{1/2} \alpha}{D_{90}} - 8.3 \right) \frac{\sin \alpha}{\tanh (k_s h)_v}$$

where S longshore transport in m³/s
D_{90} grain diameter corresponding to 90% passing (by weight)
T_s 15% excess value of wave period
H_{sd} deep water significant wave height
α angle of incidence for the waves
k_s $= \frac{2\pi}{L_s}$ where L_s is the 15% excess value of wave length
h depth of foreshore
$(k_s h)_v$ (kh) on the foreshore

A comparison with this formula is not possible because it predicts zero transport for all the test conditions except for test no 10 in which case the formula predicts the measured transport with good accuracy. One reason why the formula is not generally applicable is probably that it is based on tests with H_s/D_{n50} values outside the berm breakwater range. The implemented empirical threshold value for the onset of longshore transport $(H_{sd} \cos^{1/2} \alpha/D_{90} = 8.3)$ does not correspond to the observations in the Aalborg University tests.

A study of a general parametric representation of the test results is in progress. However, it is believed that more tests will be necessary to confirm the validity of any parametric representation of the long structure mass transport.

CONCLUSIONS

The roundhead erosion and the erosion of the trunk in oblique waves have a very strong non-linear dependency on the sea state. Below a certain sea state threshold value the erosion rates are very small, but excess of this value causes a drastic increase in the erosion. Consequently identification and consideration of this threshold value are of great importance in the design process.

ACKNOWLEDGEMENT

The help of Mr. Van der Meer in estimating cross section profiles during the planning of the tests is gratefully acknowledged.

REFERENCES

Van der Meer, J.W., Pilarczyk, K.W. *Dynamic stability of rock slopes and gravel beaches.* To be published in proc. of the 20th Int. Conf. on Coastal Engineering, Taipei, 1986.

Juul Jensen, O., Klinting, P. *Evaluation of scale effect in hydraulic models by analysis of laminar and turbulent flows.* Coastal Eng. vol. 7 nr. 4, Nov. 1983, pp 319 - 329.

Van Hijum, E., Pilarczyk, K.W. *Equilibrium profile and longshore transport of coarse material under regular and irregular wave attack.* Delft Hydraulics Laboratory. Publication no. 274, 1982.

Nielsen, Peter. *Some basic Concept of Wave Sediment Transport.* DTH, 1979.

Bijker, E.W. *Littoral Drift as Function of Waves and Current.* Proc. 11th Coastal Engr. Conf., London, 1968.

HYDRAULIC PERFORMANCE
OF
BERM BREAKWATERS

by

Ole Juul Jensen and Torben Sorensen

Abstract

Rubble mound breakwaters have been used for centuries for the protection of harbours. In many cases breakwaters were built in relatively deep water and exposed to waves too severe in relation to the size of rock used for construction. Furthermore, they were often built with a steep slope, and consequently, severe damage occurred. In some cases, breakwaters have been repaired by a continuous supply of stones until an almost stable equilibrium slope developed. In this way, the breakwaters at Cherbourg, Plymouth and Holyhead, Refs. /2/ & /3/ were developed. At certain places in nature the same may be observed for gravel beaches, where the available material by wave and tidal action is reshaped until an almost equilibrium situation occurs. In recent years, the concept of unconventional rubble mound breakwaters, i.e. berm breakwaters, has gained much attention among researchers and engineers as an economical method to build breakwaters at certain sites. At DHI, the principle of berm breakwaters was first used in 1978 for the Skopun Breakwater Extension, Faroe Islands.

Résumé

Les brise-lames en enrochements ont été utilisés depuis des siècles pour la protection des ports. Dans de nombreux cas les brise-lames étaient construits dans des eaux relativement profondes et exposés à des vagues trop fortes par rapport aux dimensions des roches utilisées pour la construction. De plus, ils ont souvent présenté une pente trop raide et ont en conséquence subi de graves dommages. Dans certains cas les brise-lames ont été réparés au moyen d'un apport continu en pierres jusqu'à ce que soit atteinte une pente d'équilibre presque stable. Les brise-lames construits à Cherbourg, Plymouth et à Holyhead (réf. /2/ et /3/) ont ainsi évolué. Dans la nature à certains endroits on peut observer le même phénomène pour les plages de gravier où les matériaux disponibles sont remaniés par l'action des vagues et des marées jusqu'à ce que soit presque atteinte une situation d'équilibre. Au cours des dernières années le concept du brise-lames en enrochement non classique c.-à-d. les brise-lames à risberme, s'est mérité une grande attention de la part des chercheurs et des ingénieurs à titre de méthode peu coûteuse de construction de brise-lames à certains emplacements. À l'Institut danois d'hydraulique le concept du brise-lames à risberme a été utilisé pour la première fois en 1978 lors du prolongement du brise-lames Skopun aux îles Féroé.

HYDRAULIC PERFORMANCE OF BERM BREAKWATERS

By Ole Juul Jensen & Torben Sørensen

Danish Hydraulic Institute
Agern Allé 5
DK-2970 Hørsholm
Denmark

1. INTRODUCTION

1.1 The Berm Breakwater Concept

Rubble mound breakwaters have been used for centuries for the protection of harbours. In many cases breakwaters were built in relatively deep water and exposed to waves too severe in relation to the size of rock used for construction. Furthermore, they were often built with a steep slope, and consequently, severe damage occurred. In some cases, breakwaters have been repaired by a continuous supply of stones until an almost stable equilibrium slope developed. In this way, the breakwaters at Cherbourg, Plymouth and Holyhead, Refs. /2/ & /3/ were developed. At certain places in nature the same may be observed for gravel beaches, where the available material by wave and tidal action is reshaped until an almost equilibrium situation occurs. In recent years, the concept of unconventional rubble mound breakwaters, i.e. berm breakwaters, has gained much attention among researchers and engineers as an economical method to built breakwaters at certain sites. At DHI, the principle of berm breakwaters was first used in 1978 for the Skopun Breakwater Extension, Faroe Islands.

1.2 Basic Principles

The basic principles of a berm breakwater may be presented as shown in Fig. 1.

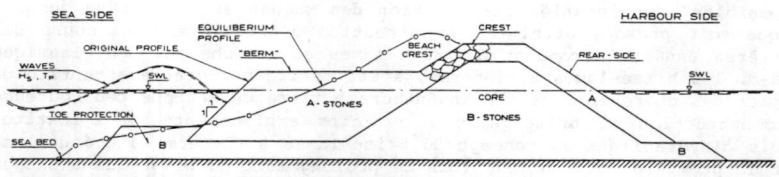

NOTE: A - STONES: LARGE SELECTED QUARRY STONES MEETING STONE WEIGHT REQUIREMENTS.
B - STONES: QUARRY RUN MEETING SPECIAL GRADATION REQUIREMENTS

Fig. 1. Principles of a Berm Breakwater.

HYDRAULIC PERFORMANCE

1. As for a conventional rubble mound breakwater, a berm breakwater requires suitable and proper toe protection if the breakwater is founded on sand in order to avoid excessive scour and sinking of the large stones into the seabed.

2. The main difference from a traditional rubble mound breakwater is with respect to the principal seaward protection. Berm breakwaters use smaller stones relative to the design wave height than traditional rubble mound breakwaters, and consequently the stones would if placed traditionally in two layers have to be placed on a very flat slope to make them stable. This would require heavy equipment, i.e. cranes with long reach and high moment capacity.

   Instead of placing the stones in two layers on flat slope for construction, they are placed in a heap on the seaward face. This requires less heavy equipment. Later the stones are reshaped by wave action until an equilibrium slope is developed. The main objective of the studies of berm breakwaters is thus to determine the necessary size and extent of the heap of stones to make sure that there is enough material in the heap to form the equilibrium slope.

3. The crest and rear side of a berm breakwater behave basically in the same way as on a traditional rubble mound breakwater. Thus, as a minimum requirement, the crest level should be high enough to prevent serious damage due to overtopping during design conditions (waves and water level). Other breakwaters may require a higher crest in order to further reduce overtopping to an acceptable level. This is especially the case if there is a reclamation behind the structure or a harbour basin sensitive to overtopping water.

4. Most berm breakwaters use fewer gradations of quarry stones than traditional rubble mound structues. It is possible (see example in Figs. 5 & 6) to limit the number of gradations to two. In this case, the small stones are used as core material and bed/toe protection, and the larger stones for the berm and armour layer on the crest and rear side.

   It is important, however, to emphasize that for this simple solution, it is of utmost importance that the large stones in the coarse gradation are carefully sorted in the quarry to make sure that this fraction contains no small stones or fines. This is important as the porosity of the large stone material ensures that the breakwater has a high wave energy absorbtion capacity.

5. The breakwater head on berm breakwaters constitute a special problem as the mode of transport of stones on the round-head is different from that on the trunk. This will be further discussed in Section 3. Normally, it will not be feasible to make the round-head on a berm breakwater with a berm profile. It is more advisable to apply either a traditional rubble mound type round-head with larger armour stones or concrete units or instead to use a type of solid breakwater head, i.e. caissons or similar.

2. THE SEAWARD PROFILE UNDER WAVE ATTACK

2.1 The Parameters of Importance

The seaward profile of a berm breakwater requires special attention. The reshaping process develops according to complex "rules" which depend on the following parameters:

1. The stones present on the slope, i.e. the average stone weight, $\bar{w}$ or the nominal diameter, D_{n50}, and the density of the stones, ρ_s. The relative stone density (stone density relative to water density) can also be expressed as $\Delta = (\rho_s/\rho_w - 1)$. According to Ref. /1/, the gradation of the stones plays a relatively minor role for the stability. Ref. /1/ defines $D_{n50} = (w_{50}/\rho_s)^{1/3}$ as the parameter for the stone sizes. Quarry stones are, however, not cubes, and DHI normally uses $\bar{w} = 0.75 \cdot \rho_s \cdot \bar{d}^3$. According to Ref. /1/, the reshaping is not very dependent upon the gradation within reasonable variations, if the material is large quarry stones.

2. The amount of material available for the reshaping process, but according to Ref. /1/ not so much the original seaward profile of the material.

3. The wave conditions:

   <u>Wave Height, H_s</u>

   The reshaping does not seem to be very dependent upon the spectral shape (Ref. /1/).

   <u>Wave Period, T_p or T_z</u>

   The wave period is as important as the wave height according to Ref. /1/. Surprisingly, it is reported that the mean period T_z is a better parameter for the wave period influence than the spectral "peak" wave period, T_p.

   <u>The Storm Duration</u>

   The storm duration, i.e. the number of waves during the storm is important for the profile development. After a certain number of waves, the profile will approach its equilibrium.

4. The water level. Different water levels due to storm surge or tides, are of importance for the reshaping process. The influence of the water level may be twofold:

   i. If the breakwater is located in shallow water with the maximum waves limited due to wave breaking, the highest and most severe waves will occur at high water level.

   ii. If the tidal range, TR, is considerable compared to the wave height, say $TR \geq 1/3\ H_s$, the tide is important as it constantly shifts the water line and thereby the zone of attack of the waves. For otherwise identical wave input, the equilibrium profile will be different for a situation with a tide compared to a situation with constant water level.

2.2 The Reshaping Process

Before describing the reshaping process, it is relevant to outline some important mechanisms about the run-up process on a rubble mound. The forces on the stones or armour units on such a slope consist of the following: 1) hydrodynamic forces (drag, lift and acceleration forces) due to the flow of water, 2) the force of gravity on the unit, 3) reaction forces from neighbouring units. A stone or unit on a slope will be displaced by the hydrodynamic forces if they are exceeding the forces trying to keep the units in place (gravitation and reaction forces). On a relatively steep traditional rubble mound breakwater with slope in the range 1:1.33 to $\sim$ 1:3.0, the armour units will always tend to move downwards on the slope if displaced from their original position. Recent model studies at DHI with measurements of the forces on armour units (Ref. /4/) have shown that the forces during run-up are generally larger than during run-down, but stones are more easily moved downwards on the slope due to gravity.

For flat slopes, 1:4.0 or flatter, the balance between the run-up/ run-down process and gravity effects changes, and the general tendency for a rubble or rip-rap slope is mostly an upwards transport of material resulting in the formation of a "beach crest", see Fig. 1.

With this knowledge, it is more clear what happens when a berm breakwater is exposed to wave action.

The stones on the seaward face can initially be placed in a heap with a large horizontal berm ending seawards in a steep slope with an angle equal to or close to the angle of repose (see Fig. 1). During wave action some of the stones are moved downwards and some upwards forming a more gentle and relatively flat almost S-shaped slope around and somewhat below SWL. This process continues, if there is enough material available for the reshaping process, until a slope is attained that is in equilibrium with the incoming wave conditions. The equilibrium slope may thus be defined as the slope where there is on average an equilibrium between the forces on the stones in the upward and downward direction.

2.3 Profile Development

The ultimate result of the wave reshaping process is thus a profile nearly in equilibrium with the incoming waves. However, for any given sea state, it will require a certain storm duration, i.e. a certain number of waves to accomplish the reshaping process of the seaward stones.

In order to make a preliminary design of a berm breakwater, it is important to have an idea about the profile development and of the equilibrium slope for any given stone size and incident wave conditions. Presently, no formula or theoretical method is available for the prediction of the equilibrium slope and the shape of the stone slope. Ref. /1/ shows an analysis of a long series of model tests of the equilibrium profile which develops on an initially infinite and straight slope of stones when exposed to wave action. Based on empirical analysis of the model tests a model was developed for prediction of the equilibrium slope for given stone material and wave conditions. However, these tests were for a straight infinite slope of stones and relatively small stone sizes relative to the incoming waves, while many practical breakwater projects deal with breaking waves and allow wave overtopping. The examples dealt with in Ref. /1/ are for H_s/D_{n50} in the range 7.4 to 21.5, i.e. very small stones relative to the wave height. The DHI-examples shown later are all for H_s/D_{n50} in the range 4.1 to 4.8. Further, the results in Ref. /1/ are for a slope of homogeneous stone material, while real berm breakwater projects normally have a minimum of two gradations of stones. It is interesting to notice that a traditional rubble mound breakwater typically has $H_s/D_{n50} \cong 3.0$ (see example below).

Example (Traditional Rubble Mound Parameter)

$K_D = 3.0$, $H_s = 4.0$m, $T_z = 8$s, $\gamma_s = 2.7$ t/m^3, $\gamma_w = 1.03$ t/m^2, $\cot\alpha = 2.0$

$$\bar{w} = \frac{\gamma_s \cdot H_s^3}{K_D \cot\alpha (\frac{\gamma_s}{\gamma_w} - 1)^3} = 6.75 \text{ t}, \quad D_{n50} = (\frac{\bar{w}}{\gamma_s})^{1/3} = (\frac{6.75}{2.7})^{1/3} = 1.36 \text{ m}$$

$$\frac{H_s}{D_{n50}} = \frac{4.0}{1.36} = 2.94 \sim \underline{3.0}$$

Ref./1/ uses a parameter

$$H_o T_o = \frac{H_s}{\Delta D_{n50}} \cdot (\frac{g}{D_{n50}})^{\frac{1}{2}} \cdot T_z$$

being a parameter for the influence of both the wave height and the wave period. The parameter is a measure of the wave height relative to the stone size multiplied with the wave length relative to the stone sizes.

Note, in this example, $H_o T_o = \frac{4.0}{1.62 \cdot 1.36} (\frac{9.81}{1.36})^{\frac{1}{2}} \cdot 8 = 39$

In order to compare the profile development, the results of model tests for four different projects are shown in Fig. 2. More details on each project are given in Figs. 4-7. The profiles shown are the ultimate profiles after completion of several test sequences as shown on the figures. Only the wave conditions for the last test sequence are reported here.

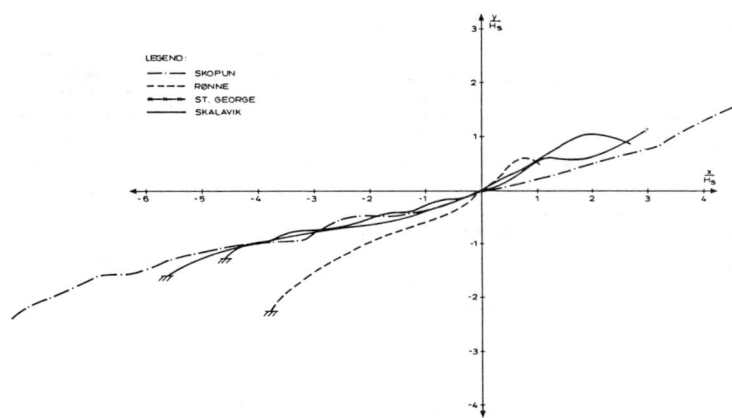

Fig. 2 Comparison of "equilibrium" profiles for four projects.

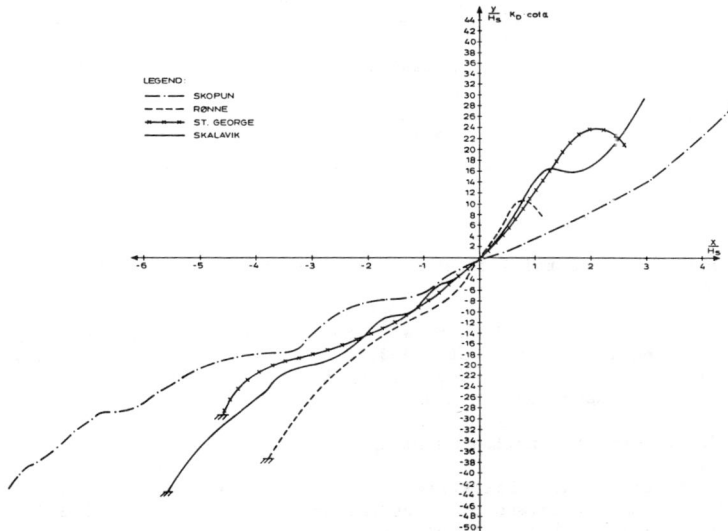

Fig. 3 Comparison of equilibrium profiles after conversion of vertical axis by multiplication with $K_D \cdot \cot\alpha$.

The details for each profile are given in Figs. 4-7 and in Table 1.

Table 1a Data for four berm breakwater projects.

Model Test Year	Project	Depth* h (m)	Stone Charact. W50 (t)	W (t)	W85/W15	D50 (m)	Wave Charact. Hs (m)	Tp (s)	Duration (h)	Steepness Hs/1.56Tp²	Stone/Waves Hs/ΔD50	Hs/ΔDn50	H0T0
1978	Skopun Faroe Isl.	21.0		12.5	10-15	1.65	7.0	18	3	0.014	4.2	2.56	78
1980	Renne Denmark	10.0	3.5		2-8	1.09	4.5	10.2	10	0.028	4.1	2.44	52
1983	St. George Alaska, USA	8.2	6.0		1-12	2.45 / 1.34	6.4	18	5	0.013	4.8	3.35	114
1986	Skalavik Faroe Isl.	15.0	8.2		3-15	2.15 / 1.45	7.0	14	2.5	0.023	4.8	2.86	73

Table 1b Results for four berm breakwater projects.

Model Test Year	Project	Equilibrium Slope Characteristics Slope SWL	Minim. Slope	Depth, DM min. slope (m)	DM/Hs	Crest Height, CH (m)	CH/Hs	Stability Coefficients K_D cotα	K_r, SWL-Slope	K_r, min. Slope
1978	Skopun Faroe Isl.	1:3.1	1:4.5	∿ 4.0	0.57	No crest	-	18.0	5.8	4.0
1980	Renne Denmark	1:1.25	1:3.3	∿ 3.0	0.67	2.8	0.62	17.0	13.6	5.1
1983	St. George Alaska, USA	1:2.4	1:6.25	∿ 4.4	0.69	6.8	1.06	22.8	9.5	5.1
1986	Skalavik Faroe Isl.	1:2.9	1:5.4	∿ 4.5	0.64	No crest	-	27.3	9.4	5.1

* Note: The depth is the actual water depth.

Comparison of the profiles in Figs. 4-7 and the data in Table 1 reveal the following:

i. The $H_s/\Delta D_{n50}$ values did not vary significantly from project to project being in the range of 2.4-3.4. By comparison with Ref. /1/, it is seen that the stones used for these four projects are relatively larger than used for the model tests.

Ref. /1/ uses the following classification:

- statically stable breakwater $H_s/\Delta D_{n50}$ = 1-4
- berm breakwaters and S-shapped profiles $H_s/\Delta D_{n50}$ = 3-6
- dynamically stable rock slopes $H_s/\Delta D_{n50}$ = 6-20
- gravel beaches $H_s/\Delta D_{n50}$ = 15-500
- sand beaches $H_s/\Delta D_{n50}$ > 300

It appears that the four projects have values close to the lower limit reported for berm breakwaters.

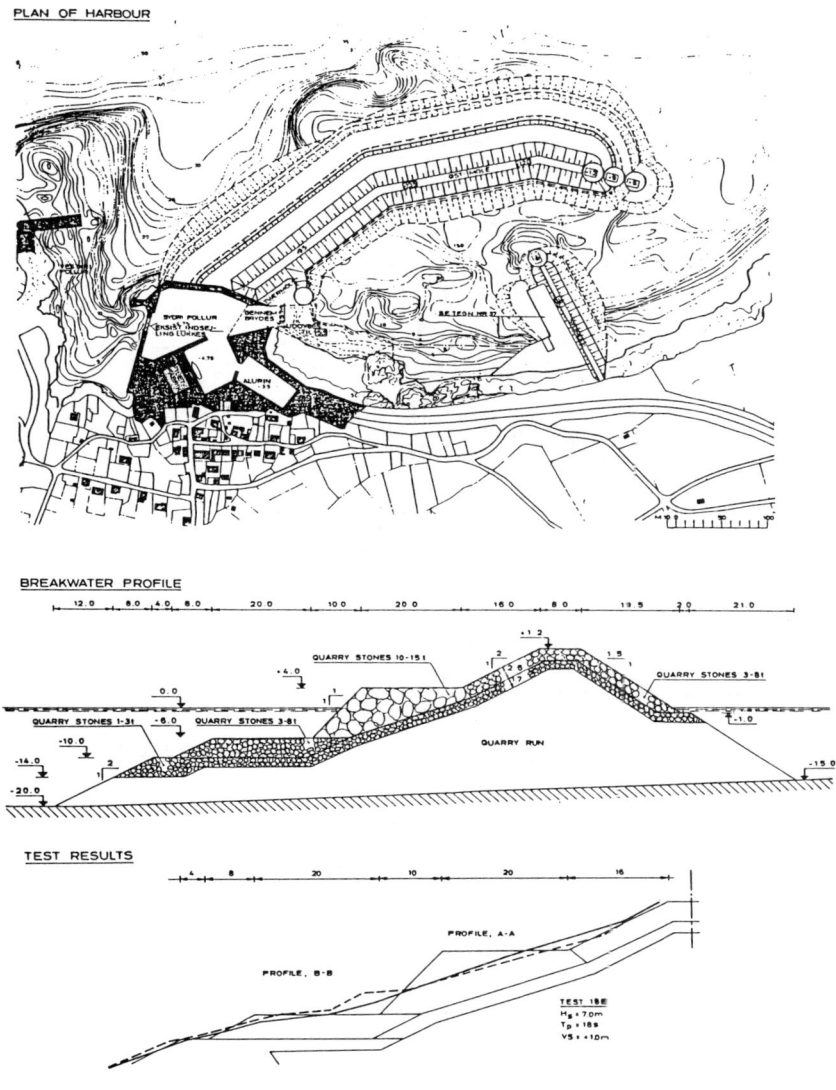

Fig. 4 Results, Skopun.

PLAN OF HARBOUR

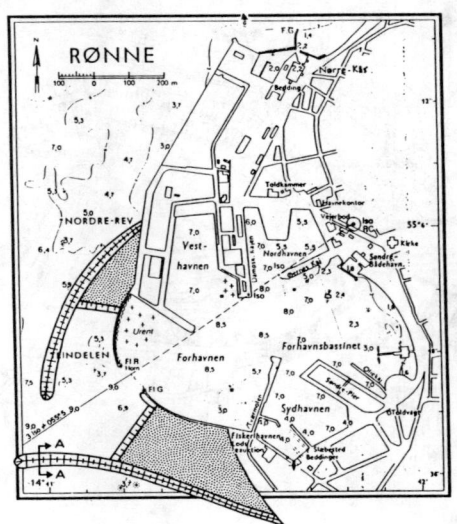

BREAKWATER PROFILE, A-A
TEST RESULTS

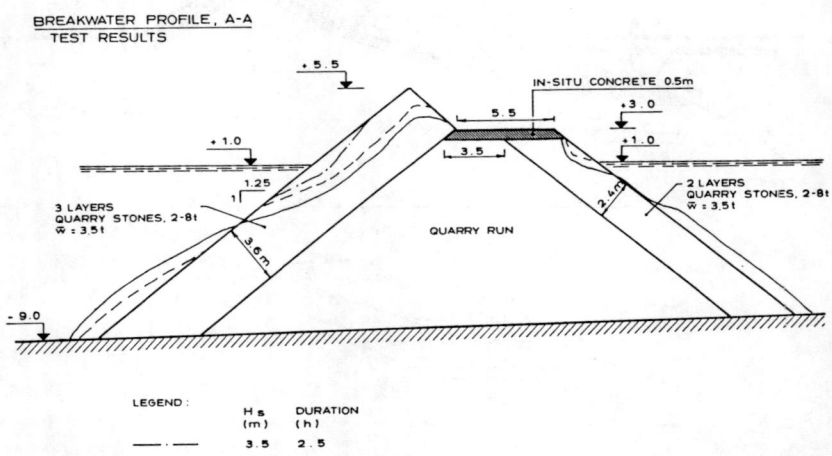

Fig. 5 Results, Rønne.

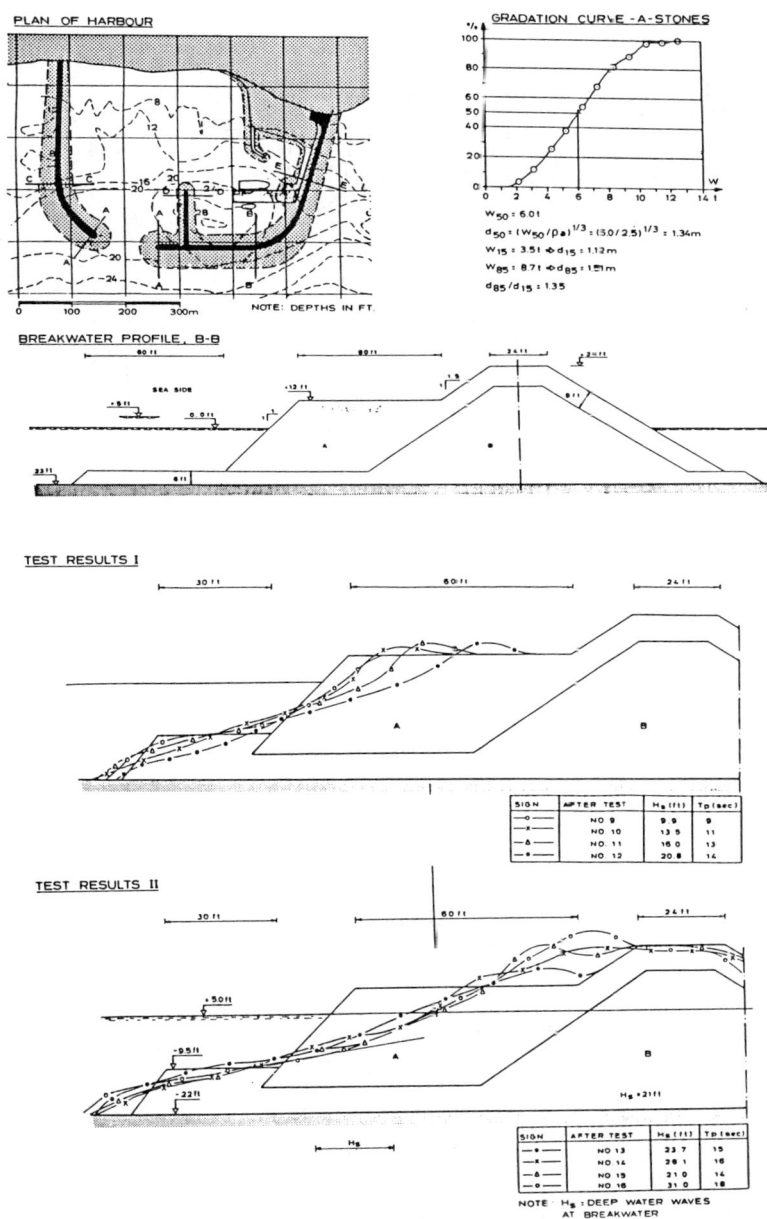

Fig. 6 Results, St. George.

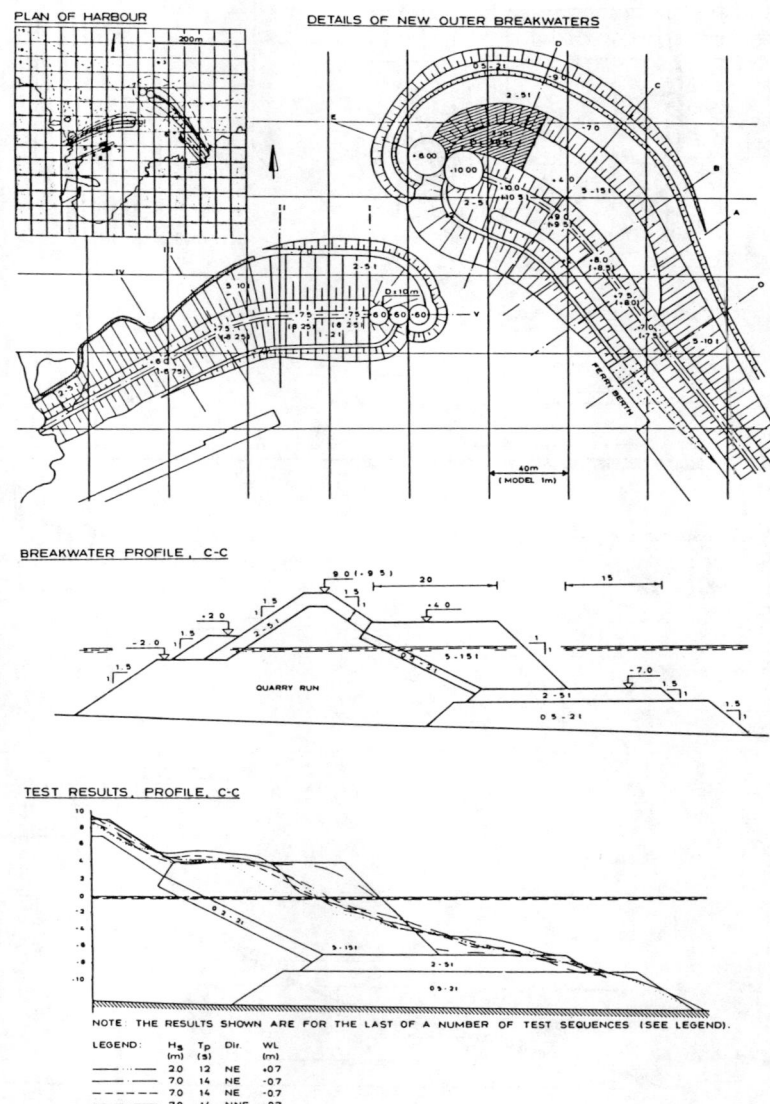

Fig. 7 Results, Skálavik.

ii. An important aspect often discussed for berm breakwaters is the possibility of "long-shore" transport of the stones if the breakwater is exposed to oblique wave attack. Not much information is available in the literature about the lower limit for "long-shore" transport of stones in slopes. Ref. /5/ presents the results of a study dealing with this aspect.

This study deals with more gentle slopes than most berm breakwaters and can consequently only be used as a first rough guideline.

Based on model testing and graphical presentation, the following empirical formula was found as the best fit to the data. $S(x)$ is the material transport. (Note: H_s = 0.10-0.21 m and $D_{90} \sim 5.8 \cdot 10^{-3}$ m).

$$\frac{S(x)}{g D_{90}^2 T_s} = 7.12 \cdot 10^{-4} \frac{H_{sd} \cos^{\frac{1}{2}} \emptyset_v}{D_{90}} \left\{ \frac{H_{sd} \cos^{\frac{1}{2}} \emptyset_v}{D_{90}} - 8.3 \right\} \frac{\sin \emptyset_v}{\tan h(k_s h)_v}$$

h is the depth and $k_s = 2\pi L_s^{-1}$ where L_s is the 15% excess value of the wave length.

It is important to notice that $S(x) = 0$ for

$$\frac{H_{sd} \cos^{\frac{1}{2}} \emptyset_v}{D_{90}} - 8.3 = 0$$

Note: According to Ref. /1/,

$H_{sd} = H_s$, transformed via the linear theory to deep water,

$\emptyset_v$ = angle of wave approach on the foreshore.

Example:

$\frac{D_{90}}{D_{50}}$ = 1.2 (corresponding to the example in Fig. 6).

Note: $\cos^{\frac{1}{2}} \emptyset_v$ is maximum for $\emptyset_v = 0$.

Thus

$$\frac{H_{sd}}{1.2 D_{50}} = \frac{8.3}{\cos^{\frac{1}{2}} \emptyset_v} \Rightarrow \frac{H_s}{D_{50}} = 8.3 \cdot 1.2 \cong 10$$

Although this is a very rough estimate of the lower limit of initiation of "long-shore" transport of material, it appears from the data for the projects, that "long-shore" transport should be insignificant for them all.

It is very important for the design of future berm breakwaters that more basic research is conducted to determine accurately the lower limit for "long-shore" transport as function of stone size, profile slope and wave conditions, etc.

iii. The four projects have H_oT_o values (see Table 1) in the range 52-114. Ref. /1/ states that for H_oT_o values smaller than 100, the slopes are stable as required in the traditional breakwater design. The formula for H_oT_o does not consider the slope of the structure. For the example given in section 2.3, H_oT_o = 100 appears high.

iv. The four profiles in Fig. 2 show some points of resemblance, especially with respect to the slope from slightly above the SWL to a distance of about H_s below SWL.

The approximate slope around SWL has been determined as shown in Table 1. It appears that the slope ranges from 1:1.25 to 1:3.1. Although Hudsons formula does not normally apply with the same stability coefficient, K_D, independent of the slope, it is interesting to notice that by using the slope found close to SWL, the corresponding K_D-factors are in the range 5.8 to 13.6, i.e. typically an average value of $K_D \cong 10$.

Somewhat below SWL, the minimum slope occurs as the material in the slope is exposed to the largest hydrodynamic forces in this region.

By the same calculation, a minimum K_D value of K_D = 4.0-5.1 is found for this region of the slope. It is interesting to notice that this value is only about 50% larger than would normally be used on a traditional rubble mound breakwater allowing for some displacements of stones during design conditions ($K_D \sim 3.0$), (see the above example). It is of further relevance that the minimum slope occurs at a distance of about 0.65 H_s below SWL. These observations are important as they can be used in a first preliminary assessment of similar berm breakwater structures.

It is important to notice that the four projects all have stones so large relative to the wave height, H_s, that the initial slope has an influence on the equilibrium slope after reshaping by waves.

Ref. /1/ states that "for $H_s/\Delta D_{n50}$ < 10-15, the initial slope has an influence on the equilibrium slope".

According to DHI's experience from other studies not described here, it seems that this limit is slightly on the high side, as model tests have shown that for $H_s/\Delta D_{n50} \simeq 6-7$, the waves are able to completely redistribute all stones on the profile.

Data from further projects and more basic research are necessary to more accurately define limits for acceptable values of the parameters of a berm breakwater to define if it has acceptable stability in the direction perpendicular to the slope. As for the "longshore" transport previously discussed, it is of utmost importance that the transport of stones up and down in the slope is at an acceptable level, once the reshaping process is completed. If "too much" movement of stones occur, there is a risk that an excessive wear of the stones will take place. Methods to quantify or estimate the wear of stones would therefore be an improvement of the technology for design of berm breakwaters.

v. As the wave action can most easily move the material on the slope from somewhat above SWL to about H_s below SWL, it is in this region that most of the stone displacement takes place and here the profiles for the four projects are most identical after reshaping. Consequently, Fig. 2 shows the largest differences in the slopes above SWL, i.e. the resultant equilibrium slope is dependent on the initial slope for the range of wave conditions and stone sizes investigated.

vi. In order to verify whether it would be possible to obtain a better comparison and in the light of the fact that the apparent stability factors after reshaping were of the same order of magnitude, an attempt has been made to draw up the profiles by multiplying the vertical axis with $K_D \cdot \cot \alpha = (\rho_s \cdot H_s^3) / (\bar{w} (\rho_s/\rho_w - 1)^3)$. The result is seen in Fig. 3, and it appears that it is mainly the Skopun profile that is different from the others. Note in Fig. 4 that the Skopun profile originally had a very wide berm (20 m) in level -6.0 m, and a flat 1:2 slope above level +4.0 m. It is assumed that these features have contributed substantially to the configuration of the equilibrium profile.

vii. The crest height of a berm breakwater can be slightly reduced compared to a traditional rubble mound breakwater because the run-up is reduced due to the flatter slope. Table 2 shows data for the four profiles in question.

It is clear from the comparison that the necessary crest elevation on a breakwater (relative to the design wave height) depends on a number of parameters, of which the slope and the wave steepness are the most important. The stone size on the rear side and the width of the crest and the slopes on both sides of the crest are important as well. It is further of relevance whether damage is permissible on the rear side of the structure. For example, in the case of Rønne where the crest is very low, the model tests have shown that damage occurred on the rear side, but that this damage did not tend to spread further. The client for this breakwater has accepted this and has easy access to stones and is already practicing regular maintenance and repair work on a similar structure.

Table 2   Crest Elevation Comparison

Project	Crest Elevation (m)	WL (m)	H_s (m)	T_p (s)	Steepness	Freeboard, Δh (m)	$\dfrac{\Delta h}{H_s}$
Skopun	+11.0	+0.70	7.0	18	0.014	10.3	1.47
Rønne	+ 3.6*	+1.0	4.5	10.2	0.028	2.6	0.58**
St. George	+7.95	+1.55	6.4	18	0.013	6.4	1.00
Skalavik	+9.50	+0.70	7.0	14	0.023	8.8	1.26

* After reshaping (see Fig. 5)
** Crest protected by concrete slab. Damage is permissible on the rear side.

3. ROUNDHEADS ON BERM BREAKWATERS

It is well known from many model studies and from practical experience that the roundhead on a rubble mound breakwater requires heavier protection than the trunk. The reason for this is simple, because the velocities in the wave rushing forward and upward are almost the same on the slope of the trunk and on the cone-shaped roundhead. On the trunk, the uprushing water works against gravity and consequently the stones are not easily moved during up-rush. (Note as previously explained for normal rubble mound breakwaters with slope 1:1.5, 1:2.0 or 1:3.0 that the subsequent down-rush is the critical phase of the wave motion on the slope.) Due to the energy loss during up-rush and in the beginning of down-rush the maximum velocity during down-rush will tend to be smaller than the maximum up-rush velocity. On the roundhead the water washes horizontally over the cone-shaped roundhead and the units protecting the roundhead are more easily moved in the tangential direction where gravity has much less stabilizing effect than on the trunk.

Model tests at DHI (Ref. /6/) have proved that for rubble mound breakwaters, it is normally necessary to increase the weight of the stones with a factor in the range 1.5 to 2.0 relative to the trunk. The factor depends primarily on the size of stones relative to the radius of curvature of the roundhead at the point of wave attack. DHI's experience further shows that for more complicated concrete units such as tetrapodes and dolos, a roundhead requires a larger increase in block weight (assuming the density is maintained) to obtain almost the same stability. Model tests have shown factors of weight increase of 2.3 for tetrapods and up to about 4.0 for dolos. Dolos seem to loose their interlocking effect when placed horizontally. This explains the severe reduction in stability of such units on roundheads. It is interesting to notice that the above observations are in line with the observations in Ref. /7/. This publication presents observations of stone and dolos stability for horizontally placed units exposed to oscillatory flow parallel to the surface. It is concluded that the stability of dolos and stones of the same weight were almost identical. (Note the density of stones is larger than the density of the concrete used for the dolos).

For the above reasons, it is DHI's opinion that for permanent roundheads berm breakwaters will normally need special roundhead protection as the berm profile used for the trunk will not be stable. Note that the fortunate situation on the trunk where the profile develops until an equilibrium situation is reached does not occur on a roundhead. If displacements occur on a roundhead, the stones will be moved backwards along the tangent of the wave direction towards the harbour or inner side of the breakwater. Here the stones are lost and have almost no stabilizing effect contrary to the situation of the trunk. On a roundhead near a harbour entrance it will normally be a requirement that the breakwater configuration is well defined and does not change with time due to depth and navigation reasons. If not very much larger stones than used on the trunk can be made available, it will be necessary to introduce other means to obtain a breakwater head with sufficient stability. In cases where a traditional rubble-mound or berm breakwater is built over more than one season, a roundhead of the berm type can sometimes be used as provisional protection in a season with rough wave conditions where the work is stopped. Whether this is acceptable, depends on the probability of serere damage and on the risk the contractor or the owner is willing to take and whether other possibilities for provisional head protection exist on the actual site. The head solutions for the four projects described in this paper are discussed in the following.

Skopun (See Fig. 4)
For the Skopun project the head consists of three cylindrical timber structures filled with stones. The sizes, foundation, levels and crest elevations appear in Fig. 4. It is important to notice that the project involved the placing of the largest stones available near the timber structures in an attempt to reduce the risk of stones damaging the timber. It is of further importance that the timber structures consist of two rings of timber and that the inner ring is connected with steel members to the outer ring. This arrangement is introduced to make sure that the outer ring does not collapse should some of the timber beams or the steel bands of the outer ring be demolished.

Rønne (See Fig. 5)
In Rønne, good quality granite stones of very large weight (up to 20 to 25 t) are easily available. Consequently the roundhead was originally planned with a traditional rubble mound roundhead. However, the breakwater which is presently under construction will for navigational reasons have a head consisting of concrete caissons.

St. George (See Fig. 6)
For the St. George project a solution for the head was developed using 33 t grooved Antifer cubic concrete units placed in the traditional manner in two layers. The slope of the armour layer was 1:2.5 on the roundhead. Wave basin model tests were performed to study the stability of the roundhead and of the transition from the berm type profile on the trunk to the conically shaped roundhead.

Skalavik (See Fig. 7)

For Skalavik the same solution as for Skopun was used. The stones in the section near the head are specified as larger than 20 t and it has been recommended that the timber beams in the zone most exposed to collision by stones should be further protected by steel pipes surrounding the timber. It is recommended as well that the Contractor makes an effort during construction to place large stones around the toe of the timber structure in such a way that they will reduce the risk of collision of stones during the reshaping process of the berm on the seaward face of the structure.

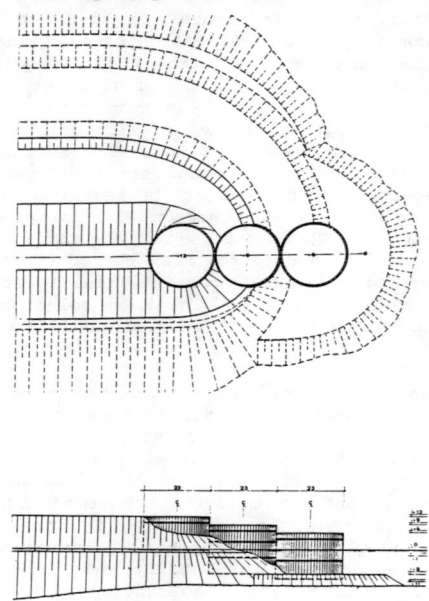

Fig. 8 Head Solution for the Skopun Project.

4. SUMMARY AND CONCLUSIONS

The paper has presented the results of four individual berm breakwater projects and used these practical examples for a more general discussion of the features of berm breakwaters and of appropriate parameters for the description of their behaviour. It is interesting to note that although the four projects discussed are very different, the comparison of the hydraulic features show great ressemblances. Berm breakwaters have attracted the attention of engineers and researchers for almost 10 years, however, the technology for the design of berm breakwaters still requires improvement, and more research is consequently encouraged. The following points are worth noting:

i. The nature of and the parameters determining the lower limit for "long-shore" transport of material on a berm breakwater is of importance and should be investigated.

ii. The possible movement upwards and downwards of stones in the equilibirum slope of a berm breakwater is of significance. More research is required to quantify this aspect and to define acceptable limits and parameters for description thereof. The associated wearing-process is an item for further research as well.

iii. The head on berm breakwaters requires more research to identify for which situations berm-type solutions may also be used for the head, with special regard to the stone size requirements.

REFERENCES

/1/ Van Der Meer, J.W. and Pilarczyk, K.W.: Dynamic Stability of Rock Slopes and Gravel Beaches. Delft Hydraulics Communication No. 379, March 1987.

/2/ Bruun, P., and Johanneson, P.: Parameters affecting the Stability of Rubble Mounds. Journ. Waterways, Harbours and Coastal Eng. Div., Am. Soc. Civ. Engs. Vol. 102, No. WW2, May 1976, 141-164.

/3/ Baird, W.F. and Hall, K.R.: The Design of Armour Systems for the Protection of Rubble Mound Breakwaters. Proc. Conf. on Breakwater Design and Construction. Inst. of Civ. Eng., London, May 1983.

/4/ Jensen, O. Juul and Juhl, J.: Results of Model Tests on 2-D Breakwater Structures. Paper to be presented at the Conf. Breakwaters '88, Eastbourne, UK, May 1988.

/5/ Van Hyum, E. and Pilarczyk, K.W.: Gravel Beaches, Equilibrium Profile and Longshore Transport of Coarse Material Under Regular and Irregular Wave Attack. Delft Hydraulics Laboratory, Publication No. 274, July 1982.

/6/ Jensen, O. Juul: A Monograph on Rubble Mound Breakwaters. Book Published by Danish Hydraulic Institute, November 1984.

/7/ Burcharth Hans F. and Thompson A.C.: Stability of Armour Units in Oscillatory Flow. Proc. Conf. on Coastal Structures 83, March 1983, Virginia, USA. Am. Soc. of Civ. Engs., New York.

APPLICATION OF COMPUTATIONAL MODEL
ON BERM BREAKWATER DESIGN

by

J.W. van der Meer

Abstract

The development of a computational model on dynamic stability is summarized. The model is able to predict profiles of slopes with an arbitrary shape under varying wave conditions. The model is used to design a berm breakwater in relatively deep water (18 m) and for severe wave conditions (H_s = 7.6 m). First the dimensions of the berm breakwater were optimized with respect to the amount of required armour stone. Then the influences of water depth, stone class and wave climate were investigated. Finally the stability after the first storms was analyzed in more detail.

Résumé

L'étude résume la mise au point d'un modéle de calcul de la stabilité dynamique. Ce modéle permet de prédire des profils de pentes avec une forme arbitraire dans des conditions de vagues variables. Le modèle est utilisé pour concevoir un brise-lames à risberme en eau relativement profonde (18 m) pour des conditions de vagues rigoureuses (H_s = 7,6m). Les dimensions du brise-lames à risberme ont d'abord été optimisées en fonction de la quantité de pierres de carapace nécessaires. Les influences de la profondeur d'eau, de la classe de pierres et du régime des vagues ont ensuite été étudiées. Enfin, la stabilité après les premières tempêtes a été analysée de façon plus détaillée.

Application of computational model on berm breakwater design

J.W. van der Meer*

Abstract

The development of a computational model on dynamic stability is summarized. The model is able to predict profiles of slopes with an arbitrary shape under varying wave conditions. The model is used to design a berm breakwater in relatively deep water (18 m) and for severe wave conditions (H_s = 7.6 m). First the dimensions of the berm breakwater were optimized with respect to the amount of required armour stone. Then the influences of water depth, stone class and wave climate were investigated. Finally the stability after the first storms was analyzed in more detail.

Dynamic stability

Most breakwaters and revetments are designed in such a way that only little damage is allowed for in the design criteria, damage being defined as the displacement of armour units. These criteria demand large and heavy rock or artificial concrete elements for armouring. A more economic solution can be a structure with smaller elements, profile development being allowed in order to reach a stable profile.

The $H_s/\Delta D_{n50}$ parameter can be used to give the relationship between different structures, see Figure 1, where: H_s = significant wave height, Δ = relative mass density and D_{n50} = nominal diameter of average stone mass.

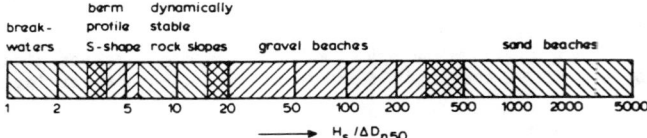

Figure 1 Type of structure as function of $H_s/\Delta D_{n50}$

Small values of $H_s/\Delta D_{n50}$ give structures with large armour units. Large values imply gravel beaches and sand beaches. Figure 1 gives the following rough classification:
- statically stable breakwaters: $H_s/\Delta D_{n50}$ = 1 - 4
- berm breakwaters and S-shaped profiles: $H_s/\Delta D_{n50}$ = 3 - 6
- dynamically stable rock slopes: $H_s/\Delta D_{n50}$ = 6 - 20
- gravel beaches: $H_s/\Delta D_{n50}$ = 15 - 500
- sand beaches: $H_s/\Delta D_{n50}$ > 300

* Delft Hydraulics, P.O. Box 152, 8300 AD Emmeloord, The Netherlands

Van der Meer and Pilarczyk (1986) described a computational model for the profile development of rock slopes and gravel beaches. The area given by $H_s/\Delta D_{n50}$ = 3-500 was covered by this computational model. The model will be described briefly and will then be applied to the design of berm breakwaters. This means that the application is focussed on the area given by $H_s/\Delta D_{n50}$ = 3-6.

Governing variables

The governing strength variables are:
stone size, grading of the stone, shape of the stone, initial slope and shape of the foreshore.

In the paper the size of armour units or gravel is referred to as the average mass of graded rubble or gravel, W_{50}, or the nominal diameter, D_{n50}, where:

$$D_{n50} = (W_{50}/\rho_a)^{1/3} \tag{1}$$

where: D_{n50} = nominal diameter (m)
W_{50} = 50% value of the mass distribution curve (kg)
ρ_a = mass density of stone (kg/m³)

The relative mass density of the stone in water can be expressed by:

$$\Delta = \rho_a/\rho - 1 \tag{2}$$

where: Δ = relative mass density (-)
ρ = mass density of water (kg/m³)

The grading of the stone is expressed here by D_{85}/D_{15}, where the subscripts refer to the 85 and 15 percent value of the sieve curve, respectively. The shape of the stone can be angular, rounded or flat. The initial profile can vary from a uniform slope to a berm profile or a structure with a low crest.

The governing load variables are:
significant wave height H_s, average wave period T_z, storm duration given by the number of waves, N, the angle of wave attack, ϕ, and water level (tide).

Conclusions on test results

One of the main conclusions of the model investigation (Van der Meer and Pilarczyk (1986)) was that in spite of different initial slopes, the same profile is reached for a large part of the total profile. This part ranges from the crest to the transition to a steep slope (the step) at the deep water end of the profile. Figure 2 gives the same profiles for a 1:5, a 1:3 and a 1:1.5 uniform initial slope. Only the upper and lower parts of the profile are in fact dependent on the initial slope. The direction of transport of material and the position of the profiles is, of course, largely influenced by the initial slope.

The same conclusion was drawn from the results of the tests in which the influence of tide was investigated. In fact, the profile changed directly with changing water level, providing that $H_s/\Delta D_{n50} > 10$ -15.

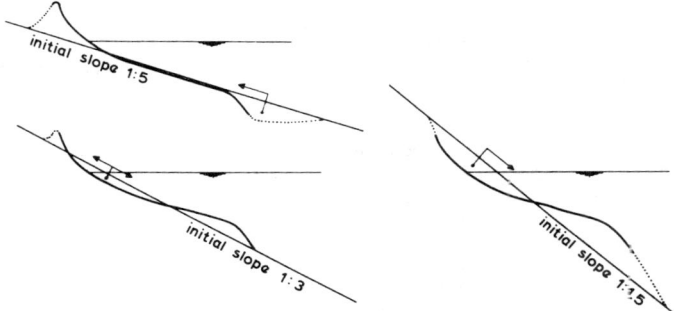

Figure 2 Profile obtained with different initial slopes

Static stability is largely dependent on the initial slope, as is clearly expressed by the well known Hudson formula. Of course, for dynamically stable structures which are almost statically stable, the initial slope has also influence on the profile. It can be stated that, for $H_s/\Delta D_{n50} < 10 - 15$ the initial slope has influence on the profile.

From the analysis it could be concluded that the wave spectrum shape had no or only minor influence on the profile, provided that the average wave period was used to compare the tests, and not the peak period. The same conclusion was found for static stability by Van der Meer and Pilarczyk (1987). The grading of the material also has no or only minor influence on the profile, using the nominal diameter, D_{n50}, as a reference.

Summarizing, from comparison of profiles it was concluded that wave height H_s, wave period T_z, number of waves, N, and nominal diameter D_{n50}, all have influence on the dynamic profile.

The spectrum shape and the grading of the material have no or only minor influence; the initial slope has no influence on a large part of the profile for $H_s/\Delta D_{n50} > 10 - 15$.

Development of a model on dynamic stability.

On the basis of the conclusions described above a model was developed to describe the dynamic profile. Two points on the profile are very important. These are shown in Figures 3 and 4, where profiles for a 1:3

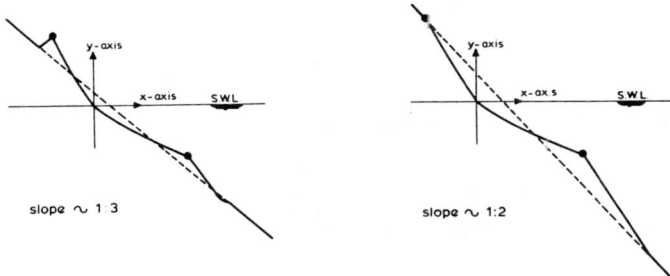

Figure 3 Schematized 1:3 profile   Figure 4 Schematized 1:2 profile

and 1:2 uniform slope are illustrated schematically. The first point is the upper point of the beach crest and the second point the transition below SWL from the gentle part to a steeper part. The local origin is chosen at the intersection of the profile with the still water level.

Figure 5 shows the model for a dynamic profile. A 1:5 uniform initial slope is shown with a high beach crest and a step. The profile is schematized by using a number of parameters all of which are related to the local origin or to the water level. The beach crest is described by the height, h_c, and the length, l_c. The transition to the step is described by the height, h_s, and the length, l_s. Curves, described by power functions, start at the local origin and go through these two points. The run-up length is described by the length, l_r. The step is described by two angles, β and γ. Finally, the transition from β to γ is described by the transition height, h_t.

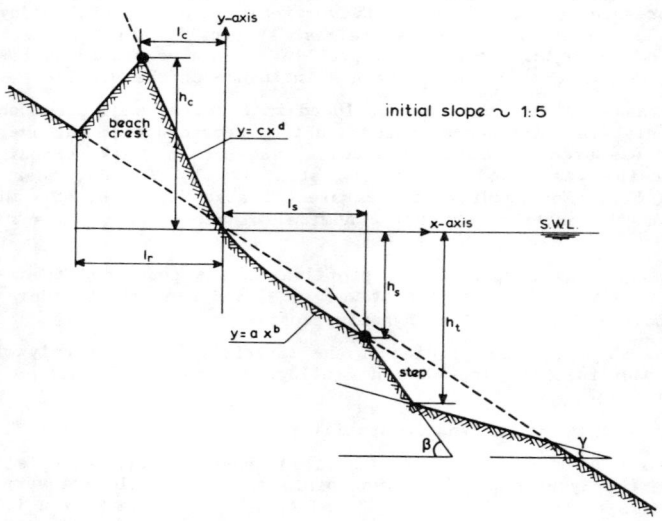

Figure 5 Model for dynamic profile

The final analysis resulted in relationships which described the above profile parameters as a function of the boundary conditions.

These relationships for the height and length parameters, the power curves and the two angles β and γ were used to develop a computer program. For low $H_s/\Delta D_{n50}$ values (smaller than 10-15) an equivalent slope angle was introduced and used in the relationships. This program can be used to calculate the profile, starting from an arbitrary slope and with varying water levels (tide) and wave conditions.

COMPUTATIONAL MODEL

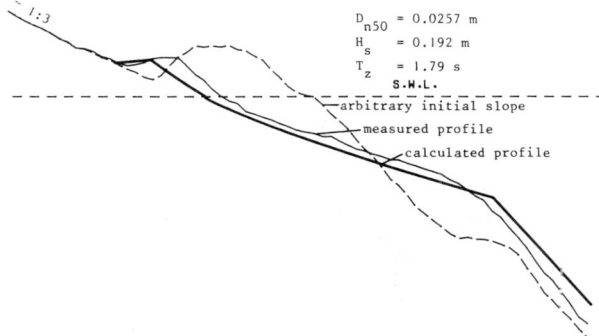

Figure 6 Measured and calculated profile developed from an arbitrary initial slope

In one test the man who constructed all the models was asked to build an arbitrary slope in the way he preferred. Figure 6 shows the slope he constructed and the measured and computed profile. The initial slope had an upper slope of 1 in 3 and a lower slope, with some irregularities, varying between 1 in 1.5 and 1 in 2. First the profile is calculated with the local origin at the intersection of the initial profile with the still water level. It is clear that this is not the right position of the profile. By means of an iteration process the profile is moved along the still water level untill the mass balance is fulfilled.

Berm breakwater concept

The berm breakwater can be regarded as an unconventional design. Displacement of armour stones in the first stage of its life time is accepted. After this displacement (profile formation) the structure will be more or less statically stable. The cross-section of a berm breakwater consists of a lower slope 1:m, a horizontal berm with a length b just above high water, and an upper slope 1:n. The lower slope is often steep and close to the natural angle of repose of the armour. This means that m is roughly between 1 and 2.

During the design of a berm breakwater the following aspects should be considered:
- Optimum dimensions of the structure: m, n, b and crest height, obtained for chosen design conditions.
- Influence of water depth.
- Influence of stone class.
- Influence of wave climate.
- Stability after first storms.

The following boundary conditions are chosen for the design of a berm breakwater which were in fact taken from a project in the Spanish Mediterenian:
- waterdepth up to 18 m.
- no tidal range
- wave climate: 1/1 year : H_s = 4.7 m T_z = 8.2 s
                1/5 years : H_s = 5.9 m T_z = 9.0 s
                1/50 years: H_s = 7.6 m T_z = 10.0 s

- storm duration of 6 hours
- available stone classes: 0.5 - 9t, D_{n50} = 1.01 m
  1 - 9t, D_{n50} = 1.11 m
  3 - 9t, D_{n50} = 1.19 m
- relative mass density: Δ = 1.55
- berm 0.5 m above the still water level (SWL).

Design of berm profile

The optimum values of m, n, b and crest height will be established for a water depth of 18 m and the 1/50 years wave conditions, H_s = 7.6 m and T_z = 10.0 s. This means that the structure is designed for $H_s/\Delta D_{n50}$ = 4.9. The optimum value of b can be established for various combinations of m and n and for the stone class with D_{n50} = 1.01 m. The criterion for the optimum value of b was the minimum value for which the upper point of the beach crest (see Figs. 3 and 4) was not a part of the erosion profile. In fact the upper point of the beach crest should lay on the initial slope, in order to prohibit erosion of the crest of the initial profile

For each combination of m and n the minimum value of b was obtained iteratively, using the computational model. Figure 7 shows the minimum lengths of the berm as a function of the upper and lower slope angles. The berm length decreases almost linear with increasing lower slope, m. The same conclusion can be drawn for the upper slope, n.

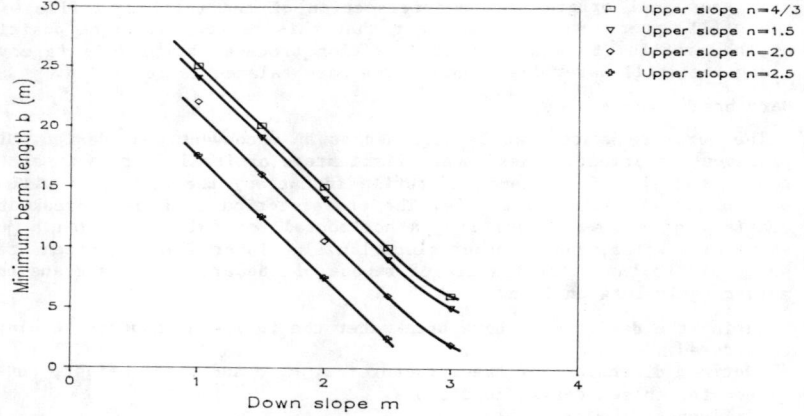

Figure 7 Minimum berm length as a function of down slope and upper slope

Figure 7 gives no information on optimum values for m and n. In fact Figure 7 gives various structures with more or less the same stability (no erosion on the upper slope). Therefore an other criterion is introduced. The amount of stones required for construction can be minimized, giving the cheapest structure. The height of the upper point of the beach crest amounted from 7 to 9 m. The crest height of the initial profile was chosen at a fixed level of 9.5 m above SWL which is about 1.25 times the significant wave height. The area of the cross-section from the crest to the toe of the structure gives a measure of the amount of stones required. This amount, B, was plotted versus the lower slope and for various upper slopes in Figure 8. The berm lengths are the same as in Figure 7.

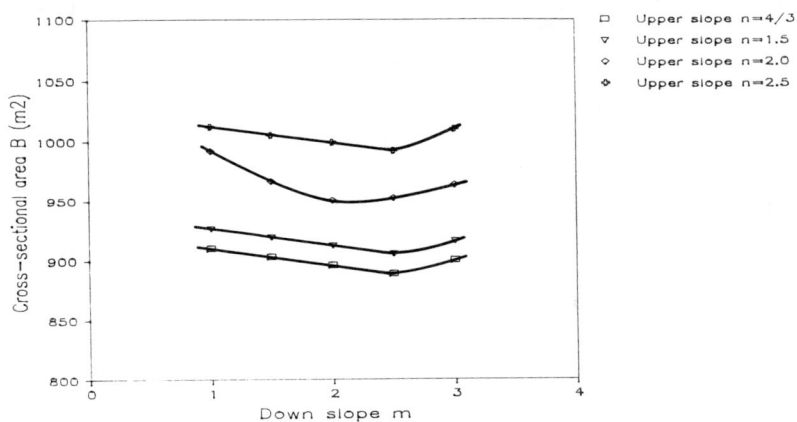

Figure 8 Cross-sectional area as a function of down slope and upper slope

From Figure 8 it can be concluded that a steeper upper slope reduces the amount of stones required. The difference is small for the steepest slopes of n = 4/3 and 1.5. The lower slope has less influence on the amount of stones required. Based on Figure 8, the lower and upper slopes were chosen for m = n = 1.5. The berm length becomes b = 19 m (Fig. 7). This choice of berm breakwater dimensions and the profile after design conditions is shown in Figure 9.

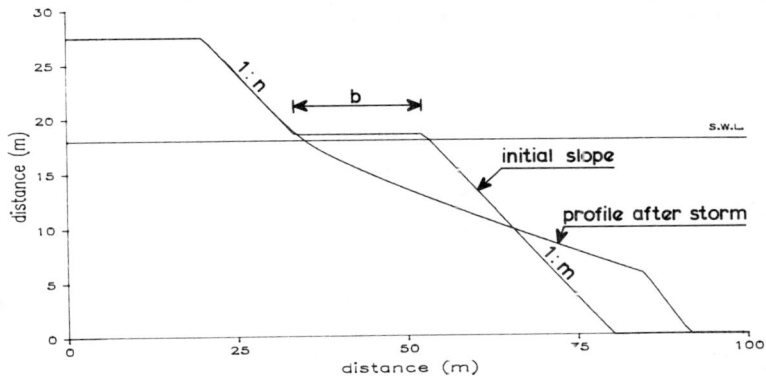

Figure 9 Berm breakwater with m = 1.5, n = 1.5, b = 19 m and profile after 1/50 years storm

Influence of water depth

With the lower and upper slopes fixed at 1:1.5 the berm length becomes 19 m for a water depth of 18 m. The berm length can be reduced in shallower water using the same design conditions. Figure 10 shows this reduction of b for shallower water.

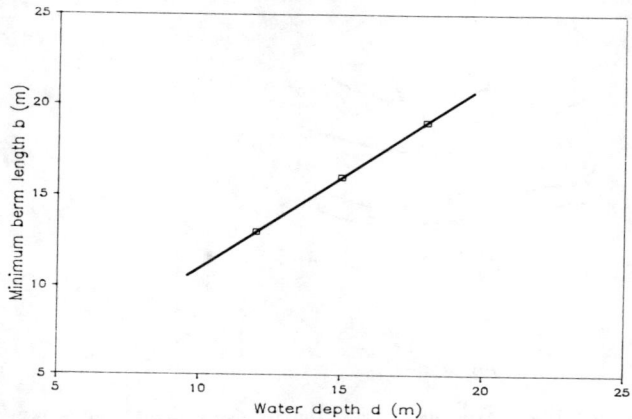

Figure 10 Influence of water depth on minimum berm length

Influence of stone class

Upto now the wide gradation of 0.5-9 t with D_{n50} = 1.01 m was used. Heavier stone will show less displacement of material. Therefore the profiles under design conditions were computed for the stone classes 1-9 t and 3-9 t according to output curves of a view quarries. Figure 11 shows all three profiles. As the differences in D_{n50} are small, the differences in profile are small too. It can be concluded that the wide (and cheaper) class of 0.5-9 t is satisfactory.

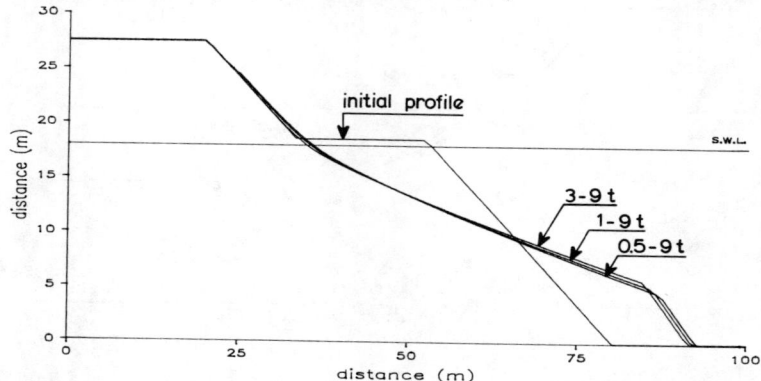

Figure 11 Influence of stone class

Influence of wave climate

The berm breakwater was designed for H_s = 7.6 m, T_z = 10.0 s and a storm duration of 6 hours, being the 1/50 years condition. The structure, however, will show profile changes for much lower wave heights. Therefore, the profile was calculated for a wave height of H_s = 4.7 m, being the 1/1 year wave height. It is furthermore interesting to know the influence of a higher wave height than the design wave height. Another profile was calculated for H_s = 9.2 m, being the 1/400 years wave height. The profiles are shown in Figure 12. The highest wave height shows some erosion of the upper slope, but under these circumstances some erosion can be allowed. The armour layer should be thick enough, however.

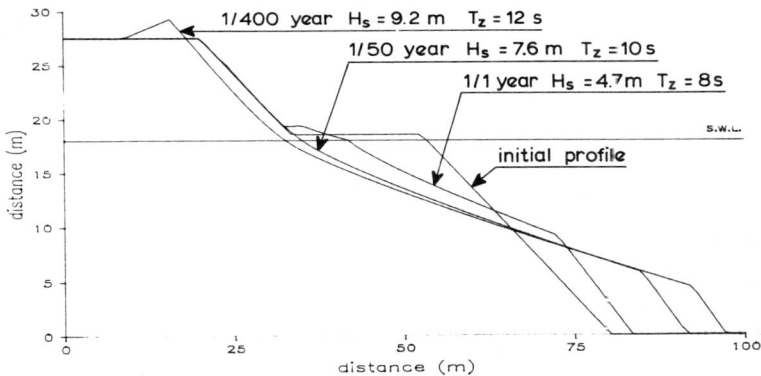

Figure 12 Influence of wave conditions

The wave period has influence on the profile and is often an uncertain factor in the design. The profiles for a lower wave period (T_z = 8 s) and a higher wave period (T_z = 12 s) were calculated and shown in Figure 13, together with the period of 10 s. The longer period gives again some erosion of the upper slope.

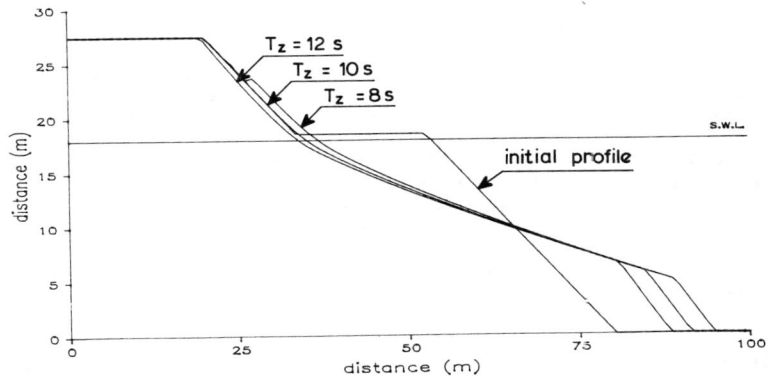

Figure 13 Influence of wave period

Finally the influence of the storm duration can be investigated. Profiles for a storm duration of 6, 12 and 24 hours are shown in Figure 14. The influence is very small. Erosion increases a little and the material is transported downwards.

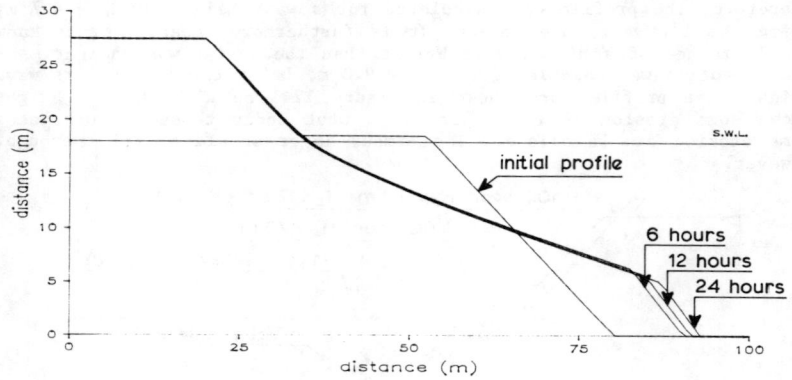

Figure 14 Influence of storm duration

Static stability

The computational model is valid for structures under dynamic stability. This means that if the wave height is too small, or the diameter too large, a profile can not be calculated with this model. The structure must then be considered as statically stable. New stability formulae (Van der Meer (1987)) can be applied in that case. The transition from dynamically to statically stable structures depends on the $H_s/\Delta D_{n50}$ value, but also on the equivalent slope angle. A gentler slope is more stable than a steep slope.

First profile changes to the berm breakwater will occur for relatively low wave heights. More severe storms will change the profile again. But how stable is the berm breakwater after its first profile changes? Consider the profile after the 1/1 year storm with H_s = 4.7 m. The profile is shown in Figure 12. The equivalent slope of the profile around the still water level is about 1:3.6. Choosing this equivalent slope, a damage curve can be drawn for static stability (Van der Meer (1987)). This curve is shown in Figure 15. Start of damage occurs for S = 2-3 and about one layer of stones is removed for S = 8-12. From Figure 15 follows that after the 1/1 years storm the structure will act as statically stable upto H_s = 6 m. In that case about one layer of armour stones will be displaced. For higher wave heights the profile will change more and will become dynamically stable again.

Other applications

The computational model can be used to describe the behaviour of rock and gravel beaches, including the influence of storm surges and tides. It can also be used to design a two-layer S-shaped breakwater. The length and slope of the gentle part of the S-shaped breakwater can be estimated and also the steeper upper and lower slopes. Another application is the prediction of the behaviour of core and filter layers under construction when a storm hits the incomplete part of a breakwater.

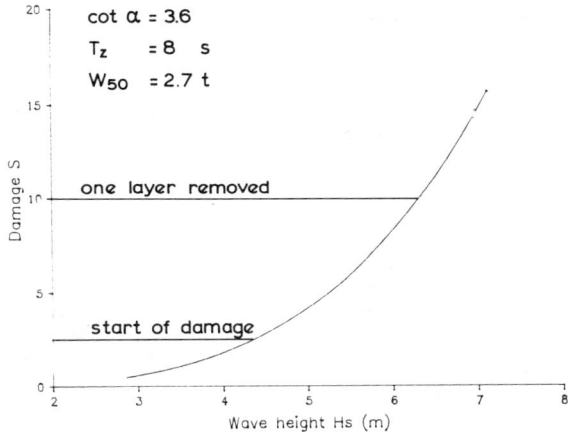

Figure 15 Damage curve for a statically stable homogeneous structure with an equivalent slope 1:3.6

Conclusions

The development of a computational model on dynamic stability has been summarized. This model was used to design a berm breakwater. The optimum dimensions of the breakwater were calculated with respect to the minimum amount of stone required for construction. The 1/50 years design conditions were used for this procedure. Influences of water depth, stone class, and wave climate were investigated. Finally the transition from dynamic to static stability was studied in more detail.

REFERENCES

Van der Meer, J.W. and Pilarczyk, K.W., 1986.
Dynamic stability of rock slopes and gravel beaches.
Proc. 20th ICCE, Taipei.
Also Delft Hydraulics Communication No. 379-1987.

Van der Meer, J.W. and Pilarczyk, K.W., 1987.
Stability of breakwater armour layers - Deterministic and probablistic design.
Delft Hydraulics Communication No. 378.

EXPERIMENTAL AND HISTORICAL VERIFICATION OF THE PERFORMANCE OF NATURALLY ARMOURING BREAKWATERS

by

Kevin R. Hall, Department of Civil Engineering
Queen's University, Kingston, Canada

Abstract

A series of hydraulic model tests were carried out to investigate the mechanism by which naturally armouring breakwaters, that is breakwaters in which the initial profile is adjusted into a more stable profile as a result of wave action, gain their stability. The breakwater models were instrumented with pressure transducers placed along the outer face of the structure and at the core/filter interface. Specially designed capacitance gauges were installed throughout the core to monitor the motion of the phreatic surface within the breakwater. A substantial reduction in the magnitude of the internal and external pressure field was found for a natural armouring breakwater compared with a conventional structure.

In addition, this paper provides a detailed summary of research on naturally armouring breakwaters which has occurred during the past 150 years.

Résumé

Un ensemble d'essais hydrauliques sur modèle a été effectué pour étudier le mécanisme par lequel les brise-lames adaptatifs, c'est-à-dire les brise-lames dont le profil initial est modifié en un profil plus stable par l'action des vagues, acquièrent leur stabilité. Les modèles de brise-lames ont été garnis de transducteurs de pression placés le long de la face extérieure de l'ouvrage et à l'interface noyau/filtre. Des jauges de capacité électrique de conception spéciale ont été installées dans le noyau afin de surveiller le déplacement de la surface phréatique à l'intérieur des brise-lames. Une réduction substantielle de l'ordre de grandeur du champ de pression, tant interne qu'externe, a été constatée dans le cas d'un brise-lame adaptatif comparé à un ouvrage classique.

Cette étude présente de plus un résumé détaillé des recherches concernant les brise-lames adaptatifs au cours des 150 dernières années.

Experimental and Historical Verification of the Performance of Naturally Armouring Breakwaters

Kevin R. Hall, Department of Civil Engineering
Queen's University, Kingston, Canada

Abstract

A series of hydraulic model tests were carried out to investigate the mechanism by which naturally armouring breakwaters, that is breakwaters in which the initial profile is adjusted into a more stable profile as a result of wave action, gain their stability. The breakwater models were instrumented with pressure transducers placed along the outer face of the structure and at the core/filter interface. Specially designed capacitance gauges were installed throughout the core to monitor the motion of the phreatic surface within the breakwater. A substantial reduction in the magnitude of the internal and external pressure field was found for a natural armouring breakwater compared with a conventional structure.

In addition, this paper provides a detailed summary of research on naturally armouring breakwaters which has occurred during the past 150 years.

Introduction

In recent years, the failure of many large conventional type breakwater structures has led to a careful examination of the physical processes of wave-structure interaction. It is known that the interaction of an incident wave with a rubblemound breakwater results in complex flow patterns involving unsteady non-uniform flow. In most cases it is desirable to construct a breakwater which works in harmony with the flow field; that is, to construct a structure with a geometry and armour stone weight gradation which results in natural profile readjustment and subsequent minimisation of the applied hydrodynamic loadings.

A naturally armouring breakwater can be described as a mound of rock, often comprised of a wide range of stone sizes, which undergoes reshaping as a result of wave-structure interaction. As a consequence of this wave action, a stable profile is developed. Two major processes occur in the development of the stable profile. First, the overall geometry of the structure responds to the nature of the hydrodynamic loadings. Material is sorted and redistributed into a profile which acts to minimise the applied forces by altering the flow field kinematics. Secondly, this natural sorting leads to consolidation (densification) of the armour layer as stones that move eventually find voids into which they nest.

This type of structure has been used extensively in the past decade and it has been found that these structures are significantly less expensive than more conventional breakwaters designed in accordance with guidelines given in design manuals such as the US Army Corps of Engineers Shore Protection Manual. The armour stones required are much smaller than those required by conventional stability formulae and a much wider gradation can be used. This allows for the design to be based on the actual quarry output rather than some pre-conceived specification for stone for which a quarry must be found. Experimental studies have indicated that the reshaped breakwater profile can be closely predicted for the design wave conditions and the available material properties. This profile can be used as an initial design. Model studies are used to optimise the design and minimise the cost.

The concept of a 'natural profile' breakwater, on which the applied forces are minimised, can be further supported by undertaking a review of breakwaters constructed during the 19th century, breakwater failures and by reviewing prototype experience with natural profile breakwaters gained in the past decade.

A vast amount of historical data is available which provides information regarding the long term performance of breakwater structures. Many of the early breakwater structures were constructed by simply dumping quarried stone, which was available at the job site, into the sea. Enough material was placed for the breakwater to eventually develop a stable profile. The natural action of the waves would redistribute and sort the stones into a stable rubblemound configuration.

Stability can be defined as either static or dynamic. Typically, in the past 50 years breakwaters have been designed to be statically stable; that is, no changes to the profile are expected after the structure is built. Alternatively, a dynamically stable structure is one which develops a stable profile following a period of wave attack at a given threshold level. A dynamically stable profile does not imply that individual particles are immovable, but rather that there is no net movement on the structure. Some particles may move around a median position.

The bulk of the research into dynamically stable systems has been undertaken for sand and gravel beaches [Kemp (1960), Watts (1964), Van Hijum (1976), Gourlay (1980), etc]. However, examination of existing breakwater structures can provide valuable data regarding the development of stable profiles of breakwaters subject to wave attack. Recently Hall et al. (1983) and Baird and Hall (1984) have presented the results of investigations conducted to produce dynamically stable breakwater designs which maximise the use of locally available quarry material.

Historical Review

Many rubblemound breakwaters built during the 19th century were designed to allow wave action to mould the structure into its final shape. Table 1 provides a summary of some of these early rubblemound

VERIFICATION OF PERFORMANCE

Breakwater	Nature of Construction	From Bottom to Near Low Water	General Slopes		
			Near Low Water	Up to High Water	Above High Water
Portrush	Loose rubble	1.75 to 1	6 to 1	3 to 1	1.3 to 1
Plymouth	Loose rubble below Hw pitched slopes above	1.67 to 1	4 to 1	5 to 1	5 to 1
Portland	Loose rubble below Hw, plumb wall above Hw	1.25 to 1	5 to 1	5 to 1	5 to 1
Cherbourg	Loose rubble	2 to 1	7 to 1	7 to 1	1.2 to 1
Alderney	Loose rubble	2 to 1	5 to 1	Plumb wall	
Kingstown	Pitched slopes of rubble	1.75 to 1	5 to 1	5 to 1	5 to 1
Holyhead	Loose rubble	1 to 1	6 to 1	6 to 1	
Cette	Loose rubble with plumb wall	1.5 to 1	3.5 to 1	Plumb wall	

TABLE 1 SUMMARY OF HISTORICAL BREAKWATERS

Table 2 Examples of Recent Mass Armoured Breakwater Designs

Location	Depth (m)	Design Wave (Height) (m)	Armour Stone Size (t)
Little Grassy, Australia	10	8*	0.002 to 10
Murray's Beach, Australia	4	3.2*	0.1 to 2
Tug Harbour, Australia	7	4.5*	0.02 to 4
Codroy, Canada	8	6.5*	0.5 to 4
North Bay, Canada	3	2.0*	0.002 to 0.7
Riviere Renard, Canada	8	6.5*	0.3 to 8
Toronto, Canada	6	4.5*	0.02 to 1.5
Roratonga, Cook Islands	5	4.1*	0.05 to 0.7
Racine, WI, USA	7	4.5*	0.3 to 3
Unalaska, AL, USA	17	11*	1.5 to 17

structures, indicating the final wave-adjusted slopes and average stone weights. Many of these structures are now into their second century of service. Figure 1 compares the relative geometries of these structures.

Perhaps the most famous 19th century breakwater, designed by John Rennie, was built at Plymouth, England. Rennie (1843) reported the results of a series of surveys of breakwater profiles he undertook in England and in France. The final design profiles for the Plymouth breakwater was based on Rennie's analysis of his field data. The structure at Plymouth has remained functional over the past 125 years, while many of its more recent counterparts have been destroyed by wave action.

Recent Research

A number of authors have conducted studies aimed at assessing profile development of sand beaches, till shorelines and rubblemound breakwater structures. These studies have been conducted in laboratories and in the field. The results of the major investigations which have direct application to rubblemound breakwater design are described below.

Popov (1960) undertook a model investigation to determine the characteristics of stable profile development on the shorelines of earth dams and reservoirs. He noted that the developed profiles were a function of the wave height and period, grain size diameter and fluid viscosity. Popov characterised the stable slopes into four distinct zones, shown in Figure 2. The equations for determining the geometry of the slope are summarised in Figure 2.

Priest et al. (1964) described the results of a model investigation in which the profiles developed as a result of wave action were measured for several breakwaters constructed with armour stones or concrete cubes. Results of their tests are summarised in Figure 3. They found that the profiles could be described by the equation:

$$Y/D = m \left(\frac{X}{D} \right)^n$$

where m = coefficient
n = exponent
X = horizontal dimension
Y = vertical dimension
D = still water depth

In their test program, the structures were initially constructed with a front slope of 1:1.5. No initial overtopping was allowed. They found that m and n were a function of a dimensionless grain diameter, d/D, (where d = average grain size), and the relative density of the material. They proposed the following equation for m and n:

$$m \text{ or } n = A \log \frac{d}{D} + B$$

In their tests they observed that once a stable section was formed,

110 BERM BREAKWATERS

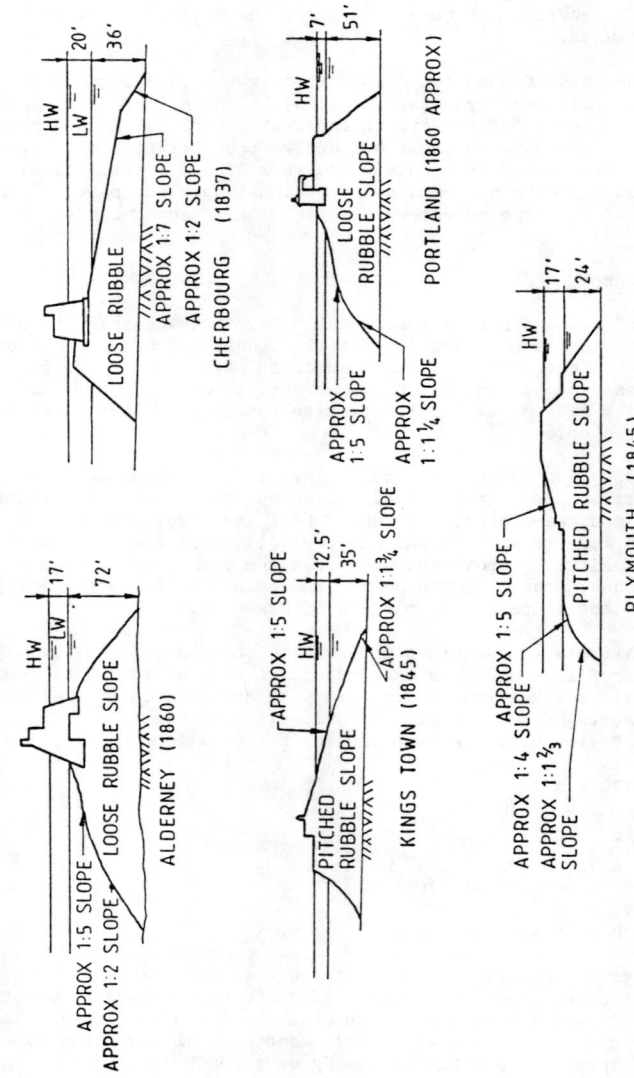

FIGURE 1 SUMMARY OF HISTORICAL BREAKWATERS

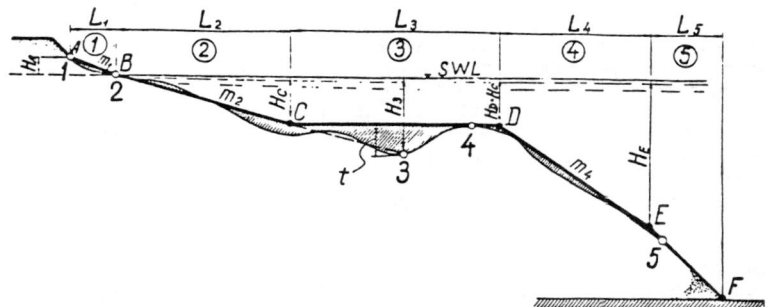

FIGURE 2   DEFINITION OF A STABLE PROFILE
(After POPOV, 1960)

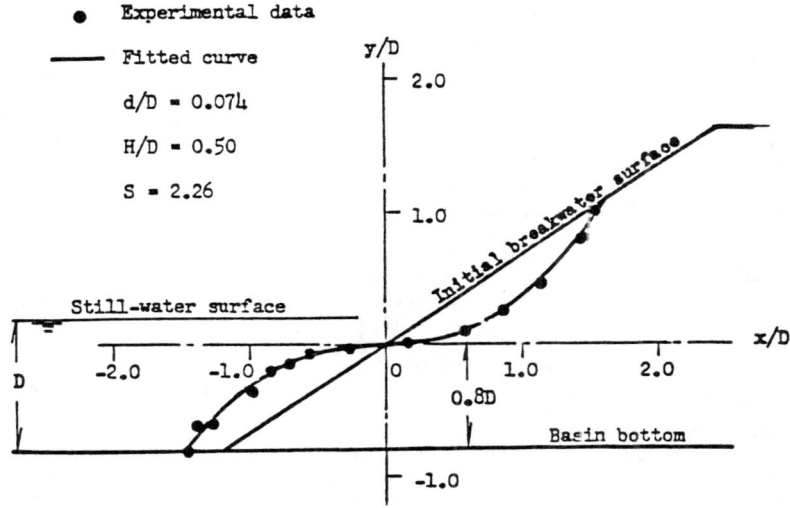

FIGURE 3   STABLE PROFILE FOR CONCRETE CUBES
(After PRIEST et al, 1964)

there was no longer movement of material, except for very lightweight material within a small zone slightly below the still water level. Hall et al. (1983), and Hall (1987) also concluded from the results of laboratory experiments that movement ceased once the stable profile was developed. The profile that had developed during early exposure to the design wave condition did not alter after a given period of time. Ahrens (1975) reported that stones which were moved by wave action tend to find stable positions which reduce the void space within the armour layer. This suggests a natural consolidation process, which is described later.

Priest et al. (1964) found that an increase in the armour stone size resulted in a decrease in the volume of material moved. The same trends were found by Hall et al. (1983) although a more complex relationship was observed which involved the relative gradation of stones.

Priest et al. found that by using concrete cubes having the same weight as stones, a decrease in the volume of material moved was observed. This suggests that the shape of the armour unit (or it's resulting permeability) has an influence on the profile development.

Kogami (1978) gave a relationship between two parameters used to describe the profiles observed during an experimental investigation of stone covered sand-mastic breakwaters. Figure 4 (a) shows the characterisation of the slope as given by Kogami. The relationship between the depth of the beach, d_c, and the deepwater wave steepness, H/L_o, is given in Figure 4 (b). Kogami found the action of stones in three regions to be distinct. In zone (a), the stones were rapidly eroded by wave action. Readjustment of the slope to a flatter value resulted in a decrease in the velocity of the wave uprush. Therefore the movement of stones eventually ceased and a stable profile was reached. In zone (b), the stone motion was found to be oscillatory with some profile redevelopment occurring. Stones from section (a) were deposited into section (c) where they were stable.

Moutzouris (1978) classified the profiles of breakwaters observed in hydraulic model tests into four distinct zones according to the type of wave action experienced by each. These zones are shown in Figure 5. Moutzouris stated that zone 1, which can be vertical, is an inert zone, that is, no significant change occurs in this zone. Zone 2 is characterised by flatter slopes and occurs in an area of wave instability, where breaking is initiated. The third zone is characterised by a horizontal section. Moutzouris stated that a water layer was formed in this zone resulting in a damping of the shock pressures resulting from the impact of the breaking waves. He stated that this zone was situated at or slightly below the still water level and was usually larger than the maximum length of the breaking zone for the largest design waves. The fourth zone was situated above the still water level and was characterised by steep slopes developed to retard wave runup.

Foster (1969) described the results of a model investigation undertaken to develop a design for a breakwater structure in Tasmania, Australia. The structure was built of quarry run obtained from

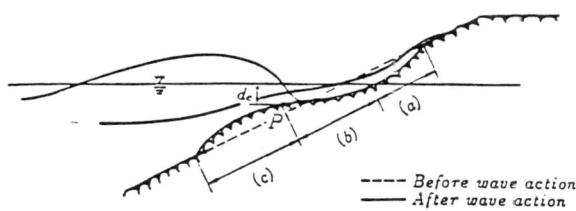

FIGURE 4(A) STABLE RUBBLEMOUND BREAKWATER CONFIGURATION
(After KOGAMI, 1978)

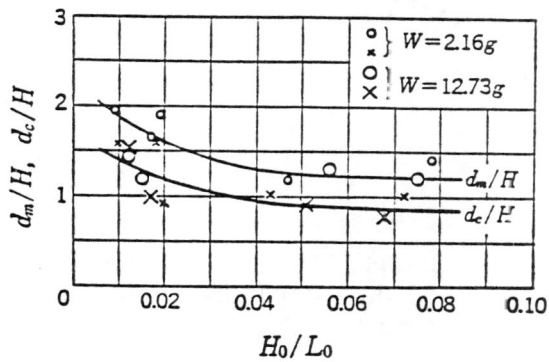

FIGURE 4(B) VARIATION OF DIMENSIONLESS CRITICAL STABILITY DEPTH WITH WAVE STEEPNESS
(After KOGAMI, 1978)

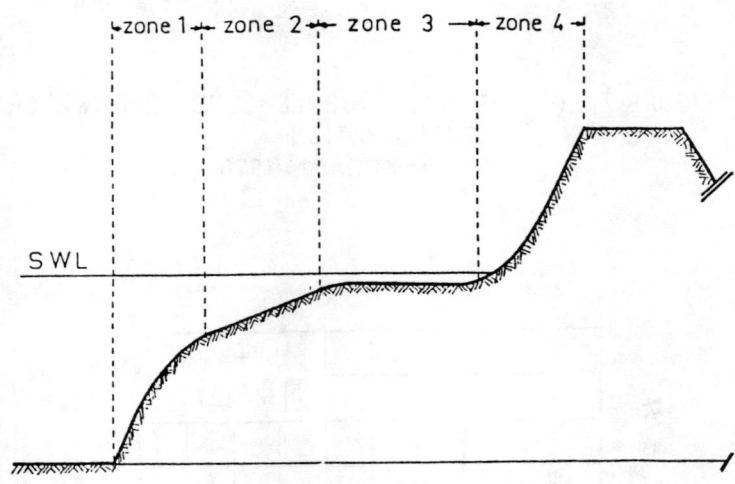

FIGURE 5 PROPOSED STABLE PROFILE FOR A RUBBLEMOUND BREAKWATER (MOUTZOURIS, 1978)

overburden and dumped at its natural angle of repose and allowed to be shaped by wave action. The material was comprised of stones weighing up to 10 tonnes maximum, but with only 5 percent greater than 2 tonnes; and 8 percent less than 2 kg. Prototype surveys undertaken 10 years following completion of the structure indicated slopes of the order of 1 in 10 through the waterline.

Van Hijum and Pilarczyk (1982) investigated the development of equilibrium profiles for coarse materials. They compared the results of regular and irregular wave tests and concluded that the resulting profiles were the same if the irregular waves were characterised by a wave height and wave period which was 15 percent higher than the deepwater significant wave height and significant wave period respectively.

The profile formation was found to be function of a number of external parameters including the wave characteristics (height, period, celerity), wave angle, viscosity, specific gravity, grain diameter, grain shape and top elevation of the beach. The shape of the equilibrium profile was not influenced by the initial profile.

They found that the grain size distribution was not constant along the equilibrium profile. At the beach crest, the diameter was found to increase with time initially, and then decrease. At the still water, a relationship between the grain diameter and the wave characteristics was not found. The diameter tended to fluctuate around the initial values. An increase in the median diameter was found between the still water level and the profile step. The depth of foreshore was found to have no influence on the equilibrium profile.

Naheer and Buslov (1983) presented two cases in which the geometrical properties of rubblemounds was investigated. The first case involved a hydraulic model study of a structure designed on the basis of a naturally armoured profile. The main armour consisted of 9-12 tonne stone. The second case involved prototype monitoring of a damaged tetrapod breakwater at Ashdod, Israel. Profiles measured on the prototype structure revealed that the original structure (1:1.33 slope) had been transformed into a naturally armoured profile. It was noted that compaction of the slope due to readjustment of the broken tetrapod pieces resulted in an increase in stability during the reshaping process. This supports data gathered by Hall et al. (1983) where nesting of individual stones within the surface layer of armour resulted in a considerable increase in the stability of the armour layer. The degree of compaction was found to be a function of the gradation of armour stone. Evidence of this consolidation suggests that the increase in stability that is associated with the natural self-adjustment of the breakwater profile cannot be replicated simply by artificial shaping (constructing the breakwater to its final profile) as suggested by Naheer and Buslov's laboratory model. Processes such as natural sorting of stones and compaction of the armour layer will not be reproduced.

Ahrens (1984) provided the results of a study undertaken to investigate 'reef type' breakwaters, which were defined by Ahrens as a homogeneous pile of stones sufficient to resist wave attack. Profiles were measured before and after testing. The final profile represented an equilibrium profile. Ahrens noted that most of the stone motion occurred during the first 10-15 minutes of testing. This is consistent with the observations of Hall et al. (1983). Ahrens observed that waves appeared to be absorbed into the structure before they could break onto it. This was due to the porous nature of the structure.

Ahrens investigated the influence of irregular waves on the performance of the reef type structure. The stability of the structures was assessed on the basis of a modified stability number, N_s, proposed by Graveson et al. (1980), where:

$$N_s^* = \frac{(H_s^2 L_p)^{1/3}}{(w_{50}/w_r)^{1/3} (s_r - 1)}$$

where N_s^* = spectral stability number
H_s = significant wave height
L_p = deep water wave length at peak frequency
w_{50} = median stone weight
w_r = unit weight of stone
S_r = relative density of stone

The spectral stability number takes into account the period of the peak energy density of the spectrum. Ahrens noted that it was convenient to relate the damage to the zero moment wave height due to the statistical stability of that parameter and because of its relation to the area under the spectrum. However, it was commonly observed that the larger waves in the distribution moved most of the stone. Ahrens concluded that the severity of irregular wave attack was measured better by the parameter $H_s^2 L_p$ than H_s^3.

Ahrens found that the threshold of stone movement occurred at a value of N_s^* of 7. For $N_s^* < 6$ little or no stone movement occurred; whereas, for N_s^* values in excess of 8, significant damage occurred.

Hall et al. (1983), Baird and Hall (1984) and Hall and Foster (1987) described a similar type of breakwater concept which has been utilised recently. Table 2 summarises the wave climate, stone gradation and water depth for some of the installations of this natural armouring type breakwater. In most cases, cost savings of 60 to 70 percent of the cost of a conventional multilayer design were achieved. The concept has the advantage of utilising the total yield of the locally available quarry. The available material is determined before the structure is designed (using trial blasting and geotechnical investigation). The structure is then optimised for the given design conditions (wave height, wave period, water depth). The geometry of the structure is determined to minimise placed quantities. The material is not simply placed in a mound at the natural angle of repose.

The final stable profile developed as a result of wave action will be influenced by a number of variables including the gradation of stones used in the armour layer, the shape of the stones and the angle of

VERIFICATION OF PERFORMANCE

wave approach. When a wide gradation of armour stone is used, a natural sorting of stones occurs. In addition, interlocking between stones is developed since stones of a given size tend to roll about on the surface until they find voids into which they nest. This allows maximum interlocking to be achieved.

The shape of the stone influences its rolling resistance, its frictional resistance and also influences the magnitude of the applied hydrodynamic forces. Both the stone gradation and the stone shape affect the porosity and permeability of the structure. This has been shown to affect the stability of the structure, by modifying the external and internal flow field kinematics [McCorquodale and Hannoura (1978)].

The structure also tends to reach its equilibrium profile at different rates for differing wave conditions. Therefore simulation of possible wave conditions during the passing of the design storm is desirable in model tests.

The angle of wave approach affects the magnitude of the applied hydrodynamic forces and therefore influences the relative distance moved by each particle in the direction of the applied force. For very lightweight material, the path of the movement of a particle is affected such that motions perpendicular to the longitudinal axis of the breakwater may be decreased while motions in the direction of the longitudinal axis are increased.

The mass armoured type of structure proposed by Hall and Foster (1987) develops its stability due to an increase in the total volume of voids of the armour layer (over conventional structures) and due to a natural sorting of stones which results in profile adjustment and the development of an armoured profile. These phenomena can be described as follows:

(i) The high total volume of voids of the berm provides a suitable medium through which the incoming wave energy can be dissipated. In a conventional breakwater, the flow is channeled up through the armour stone layer, which is usually only two units thick, and consequently the individual drag force acting on each stone is very high due to a relatively high velocity field in the armour layer [see Figure 6 (a)]. In a mass armoured concept, the incoming wave can spread its energy over a larger mass and area, thereby reducing the velocity field which results in a reduction in the forces acting on individual stones [Figure 6 (b)]. Because the forces are smaller, smaller sized armour stones can be used. The stability of the armour layer is a function of the stone gradation selected and the width of the horizontal berm (i.e. minimum thickness of armour over the core). As the stone sizes decrease to the point where individual stones are hydraulically unstable, the breakwater derives its stability through another mechanism which is described below in (ii).

(ii) Movement of the stones and consolidation of the structure increases the armour layer's stability and resistance to wave

118   BERM BREAKWATERS

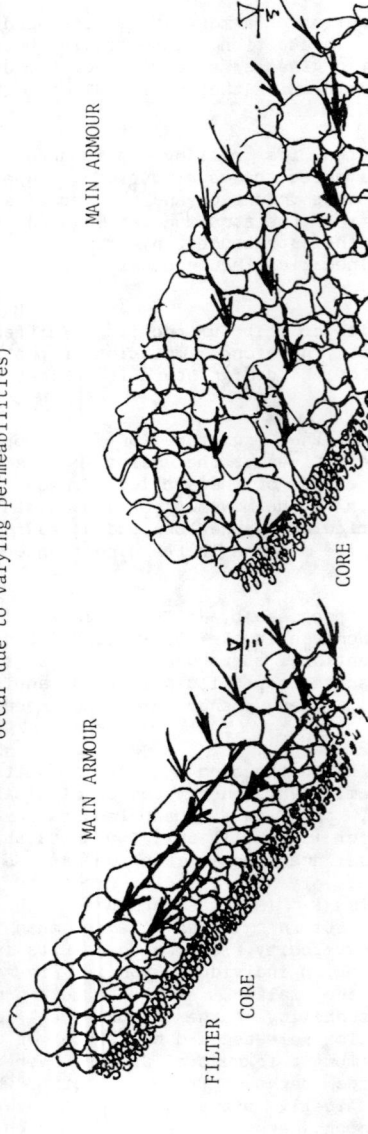

* for given wave conditions, the quantity of water entering both armour layers will be approximately the same (differences will occur due to varying permeabilities)

A) FLOW IN A CONVENTIONAL STRUCTURE

THE MAIN ARMOUR IS VERY PERMEABLE COMPARED TO THE FILTER AND CORE AND THEREFORE IS ABLE TO CONDUCT MORE FLUID WITH LESS RESISTANCE. THEREFORE, THE MAJORITY OF FLOW FROM AN INCOMING WAVE WILL TRAVEL UP THROUGH THE ARMOUR LAYER, VIRTUALLY PARALLEL TO THE SLOPE (LITTLE INFILTRATION). THE BULK FLOW AREA IS CONSIDERABLY SMALLER THAN IN THE MASS ARMOURED CONCEPT, RESULTING IN HIGH VELOCITIES AND FORCES ACTING ON INDIVIDUAL STONES.

B) FLOW IN A MASS ARMOURED STRUCTURE

THE INCOMING WAVE DISSIPATES IT ENERGY ACROSS THE ENTIRE ARMOUR LAYER WHICH HAS A MUCH LARGER BULK FLOW AREA THAN THE ARMOUR LAYER OF THE CONVENTIONAL CONCEPT. THIS RESULTS IN LOWER FLOW VELOCITIES WITHIN THE POROUS MEDIA; CONSEQUENTLY, FORCES ACTING ON INDIVIDUAL STONES ARE SMALLER

FIGURE 6   COMPARISON OF THE FLOW REGIME FOR MASS ARMOURED AND CONVENTIONAL BREAKWATERS

action. The profile of the structure is adjusted so that the applied hydrodynamic forces are reduced. The movement of stones also results in a sorting and nesting of the armour stones leading to an increase in the shear strength of the structure. As a result of armour stones nesting, the interlocking between individual stones is maximised. The natural sorting, or armouring process, involves a significant amount of motion of stones in the surface layers of the armour.

It is important to recognise that the motion observed and the reshaping of the profile do not constitute a failure. On the contrary, these processes create a surface layer of armour stones which is more resistant to wave attack than loosely placed stones that are many times the weight of the stones used in the this concept. The overall stability of the structure is improved once the nesting of stones has occurred.

Although the final shape of the cross-section may be similar to the naturally armoured profiles reported by other investigators [Naheer and Buslov (1983), Bruun and Johanneson (1974)], there are several differences. The structure does not change profile below a certain depth. Stones are not rolled out of the top of the armour layer and carried down to the seabed. The armour is consolidated because the stones in the surface layers of armour that move eventually find voids into which they nest. Also, the profile of the structure is quite regular through the water line, typically of the order of 1:5, whereas naturally armoured profiles reported by other investigators tend to have a degree of curvature to them in the region of the waterline (and thus have been described as "S" shaped profiles in some publications).

The performance of a mass armoured breakwater when exposed to wave conditions exceeding the design conditions is significantly better than the performance of conventional designs. A conventional design will lose its armour protection relatively quickly when design conditions are exceeded and this may lead to breaching of the breakwater. Damage to the mass armoured design takes place at a much slower rate because the incident wave energy is dissipated over the armour layer and the energy available to remove material from the upper section of the structure is considerably reduced.

Summary of Previous Research

A number of studies directly applicable to determining the stable breakwater profile for a given wave climate, stone size, relative stone gradation, and initial breakwater geometry have been summarised. A comparison of some of these profiles is presented in Figure 7. The plot shows non-dimensional horizontal and vertical (X and Y) coordinates referenced to the intersection of the stable profile and the mean water level. The profiles were made dimensionless by multiplying the X and Y offsets by $(M/\rho)^{1/3}$, where

M = average armour mass

ρ = density of the individual armour units

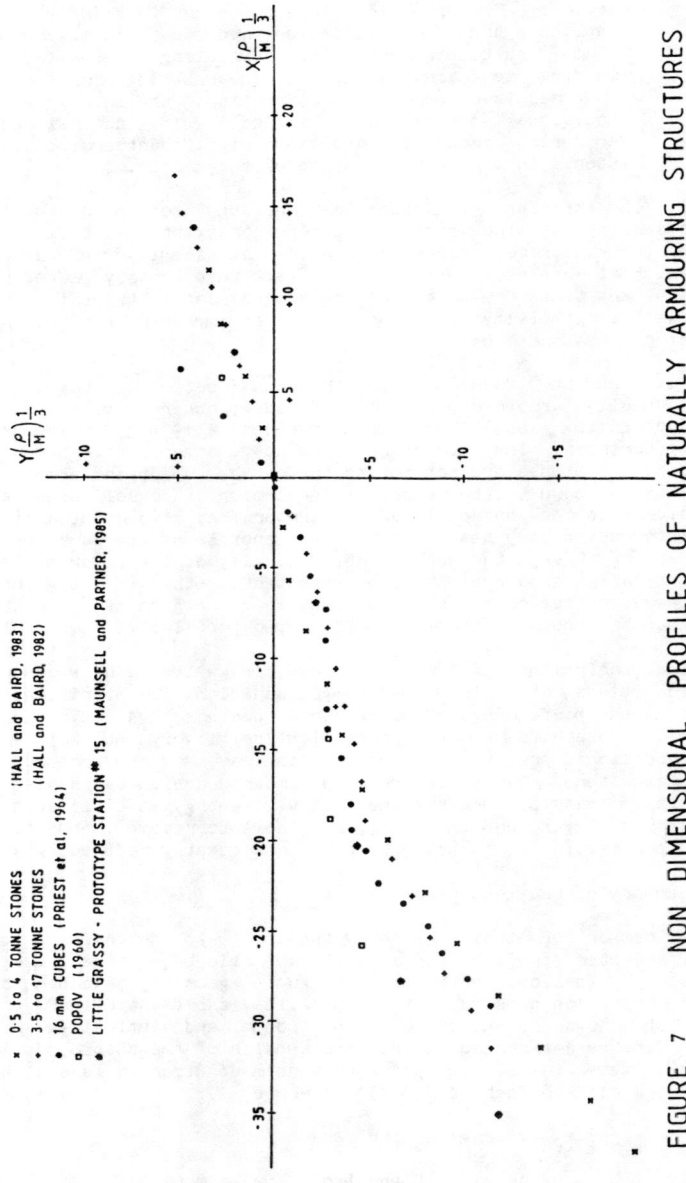

FIGURE 7   NON DIMENSIONAL PROFILES OF NATURALLY ARMOURING STRUCTURES

The resulting dimensionless profiles exhibit a distinct similarity, even though the profiles range from clay size particles to massive armour stones placed in a variety of initial geometries and subjected to a wide variation in incident wave climate.

From the results presented in Figure 7 and recalling the nature of the applied hydrodynamic forces, a number of general trends listed below may be deduced. These trends allow for further support to be developed for proposing the widespread use of naturally armouring rubblemound breakwaters.

(i)   Flatter slopes develop in areas where the hydrodynamic forces are maximum on conventional structures; that is, in the area just below the still water level. The slope readjustment reduces the magnitude of the forces in this area. This results in the ability to use smaller stones.

(ii) Below a certain elevation, no significant profile adjustment occurs, indicating an area in which the hydrodynamic forces are relatively small. Smaller size stones can be used in this area.

(iii) In the area above the still water level, the hydrodynamic forces lessen as the elevation increases; therefore a return to steeper slopes may be achieved in this area (for a given stone size). Alternatively, if the stone size is reduced, flatter slopes will prevail in this area as well as in the area identified in (i).

By observing the trends described above, one can conclude that a stable breakwater can be achieved by utilising a wide range in armour stone sizes, departing from a conventional design. Due to the complex nature of wave-structure interaction, the intensity of the applied forces varies with location on the structure. Armour stability can be achieved by various methods. In most cases, the most economical design is one in which the available material is placed to the geometry dictated by the applied forces. The most obvious way of achieving this goal is to deposit materials in sufficient quantities and let the waves move the material (natural sorting) to the final slopes and depths required to achieve stability.

Alternatively, by understanding the general trends and by conducting the proper investigations, the geometry of a stable structure for a given material and wave climate can be determined and the structure could be constructed as close as possible to the approximate stable profile, thereby reducing the amount of readjustment necessary and the time required to achieve the stable profile. However, by constructing a breakwater in this manner, the additional stability due to the natural armouring that occurs as a result of stone motion induced by wave attack would not be developed.

Experimental Studies

Introduction

A comprehensive set of experimental studies was undertaken using the facilities of the Water Research Laboratory of The University of New South Wales, (Hall, 1987). The studies were undertaken primarily to investigate the mechanism of wave energy dissipation that occurs throughout the various zones (core, filter and armour) of a rubblemound breakwater subjected to monochromatic wave attack. In addition, information regarding the phreatic surface motion within the core and filter of the structure was collected. The influence of the armour unit type, relative geometry of the armour layer, breakwater slope and the material used to construct the various layers of the structure on wave energy dissipation, wave runup and rundown on the outer surface of the structure and the internal flow generated within the structure was assessed.

These studies were undertaken in a two dimensional wave flume in which a rubblemound breakwater, instrumented with pressure transducers and capacitance gauges, was subjected to monochromatic wave attack. The rubblemound structure consisted of a core (3.5 mm angular stone), a filter layer (16 mm angular stone) and an armour layer (consisting of 1 to 5 layers of 50 mm stone).

Preliminary work was required to determine the flow resistance characteristics of the media used to construct the breakwater test section. This work was undertaken in a steady flow permeameter.

Experimental Setup

Tests were undertaken using two types of breakwaters; a conventional multi-layered design and a natural armouring structure. The profile of the natural armouring (or natural profile) test structure was selected by using the dimensionless horizontal and vertical co-ordinates for this type of breakwater, presented in Figure 7, and adapting the non-dimensional profile for the water depth and armour stone size used in this study. A comparison of these structures is presented in Figure 8.

Both types of structures were instrumented in the same manner. Each structure was instrumented with 10 pressure transducers, five on the outer slope of the structure at elevations of -20, -10, two at 0 and +10 cm (all elevations referenced to the still water level) and five on the outer slope of the core at elevations of -20, -10, -5, 0 and +10 cm placed along the centreline of the breakwater. Five capacitance gauges for measuring the internal water surface movement were located along the centreline of the structure.

Figure 9 shows the layout of the instrumentation for the natural armouring structure.

All experiments were conducted in a 0.9 m wide wave flume which has an overall length of 50 m and an overall depth of 1.75 m. The breakwater models were located approximately 35 m from the face of the

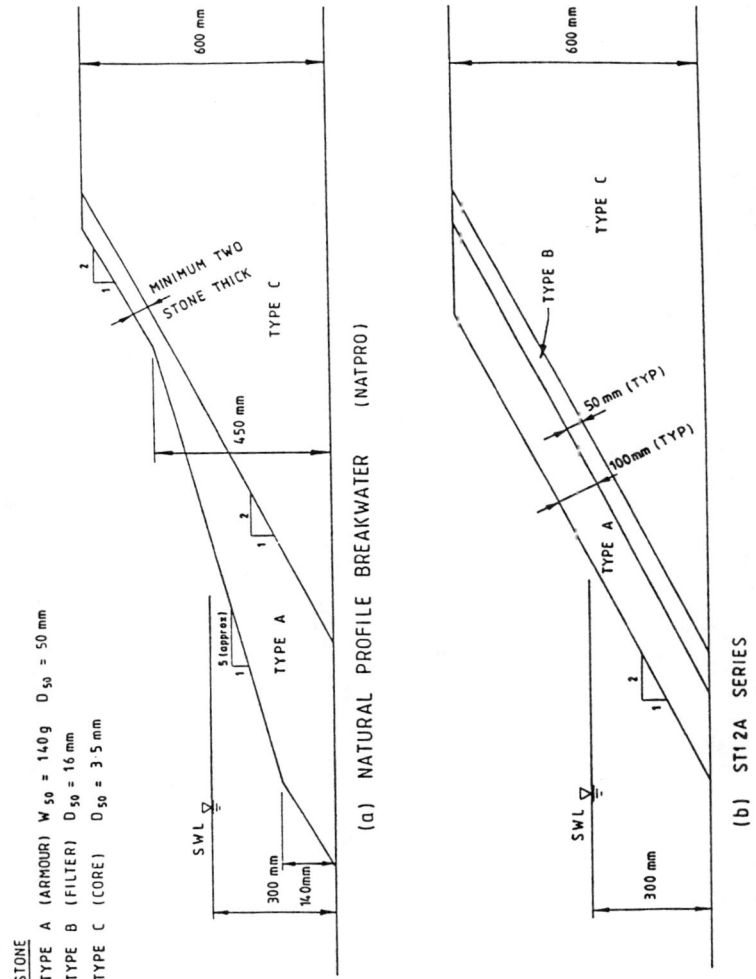

FIGURE 8   PHYSICAL RELATIONSHIP BETWEEN NATPRO AND ST12A SERIES BREAKWATER

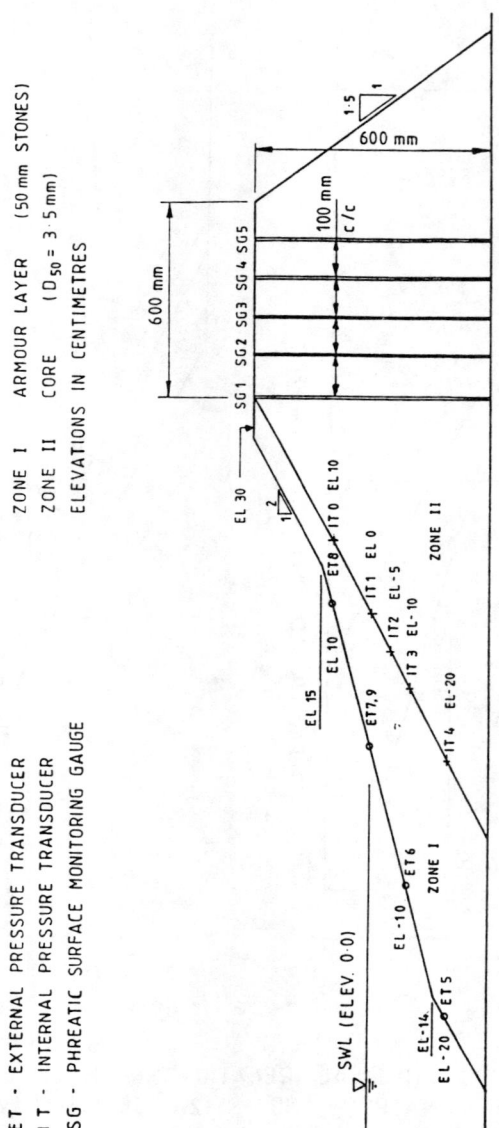

FIGURE 9   CROSS SECTION AND INSTRUMENT LAYOUT FOR NATURAL PROFILE BREAKWATER

wave paddle and were placed on a false floor, 1 m above the floor of the flume. This was done to maintain a suitable depth of water at the wave paddle to enable generation of the desired wave heights.

All tests were undertaken with the still water level located 30 cm above the flume bottom. Fifteen combinations of wave height and wave period were used for each test structure. The waves ranged in height from 30 to 200 mm and in period from 0.8 to 2.0 seconds (giving a corresponding range of surf similarity parameter of 1 to 10).

The pressure field on the outer slope of the structure, resulting from the interaction of the incident waves with the structure, and the pressure field at the filter/core interface were measured using a set of miniature stainless steel diaphragm pressure transducers. The transducer type used to measure both the external and internal pressure field was the Data Instrument AB-15-psig transducer, capable of measuring gauge pressures ranging from 0 to 100 kPa.

The external transducers were mounted inside a restraining device which was designed to maintain the porosity of the armour layer in order to ensure minimal interference of the internal flow.

The internal transducers were mounted in an aluminium frame which was traversed along two stainless steel tracks to adjust the transducer spacing to suit the breakwater slope. The transducers were located at the core/filter interface at elevations -20, -10, 5, 0 and +10 cm.

The water level in the core of the test structure was monitored on a 100 mm grid spacing across the breakwater crest using capacitance type gauges. Each gauge consisted of a capacitance wire mounted within an 11 mm diameter polycarbonate tube. The tube was drilled in a pattern to simulate, as well as possible, the porosity of the core material, thereby minimising the influence of the gauges on the flow within the core.

Automated acquisition of data provided by the pressure transducers and water level gauges was accomplished by using a Compaq Portable Microcomputer linked to a Metrabyte "Dash-16" analogue to digital conversion board. The microcomputer had a fixed disk storage capacity of 10 Megabytes.

Sixteen channels of data were recorded simultaneously at sampling rates of up to 1000 samples per second per channel, requiring a large data handling capability. The 16 channels consisted of 10 pressure transducers, 5 internal water level gauges and 1 external water level gauge.

Figure 10 shows a schematic layout of the data acquisition system.

The materials used to construct the test structures are summarized below:

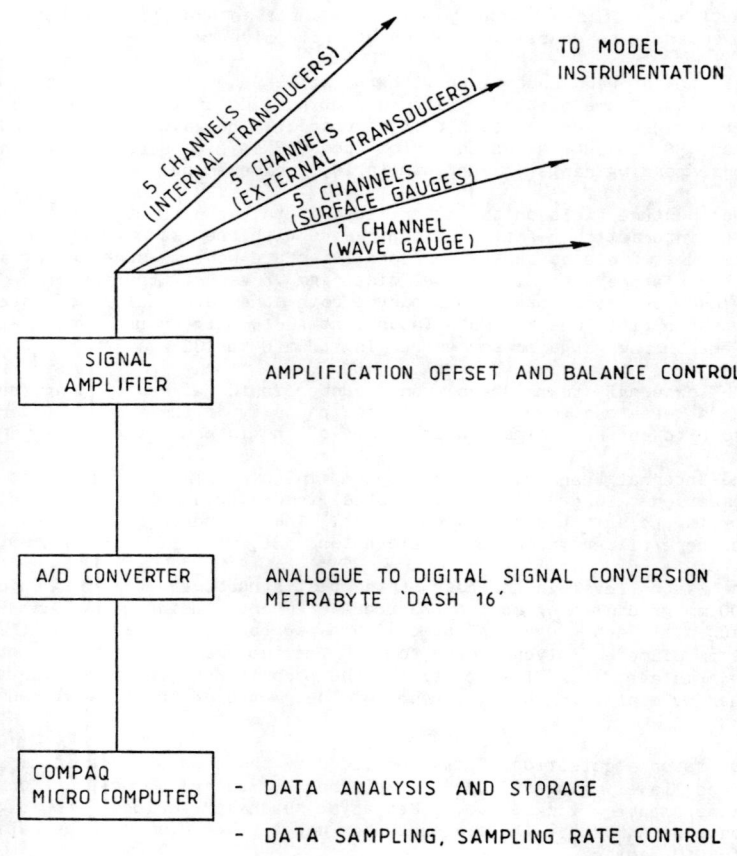

FIGURE 10   SCHEMATIC LAYOUT OF DATA ACQUISITION SYSTEM

Armour	$M_{50} = 121$ g	$D_{min} = 32$ mm	$S_r = 2.65$
	$M_{min} = 80$ g	$D_{50} = 40$ mm	
	$M_{max} = 180$ g	$D_{max} = 50$ mm	
Filter	$D_{min} = 2$ mm	$S_r = 2.65$	
	$D_{50} = 16$ mm		
	$D_{max} = 20$ mm		
Core	$D_{min} = 1$ mm	$S_r = 2.62$	
	$D_{50} = 3.5$ mm		
	$D_{max} = 8$ mm		

Test Results

(1) External Pressure Measurements

(a) Natural Profile Results

The physical layout of the natural profile breakwater was shown in Figure 9. The results of the external pressure measurements obtained on the outer slope of this profile are described in this section. The wave pressure acting on the external slope of the breakwater was measured at four elevations, +10 cm, 0, -10 cm and -20 cm (referenced to the still water level). The pressure was measured at a sampling rate of 50 Hz for a total recording length of 10 wave periods. The data was subsequently phase-averaged so that each test segment (wave height - wave period combination) was characterised by a time series having a record length equal to the wave period. For the purpose of further analysis, each phase-averaged time series was characterised by a maximum differential pressure head, ΔP, which describes the maximum variation in recorded pressure over the period of the wave. All pressure measurements were reported in terms of the equivalent head of water. The ratio of maximum differential pressure head, ΔP, to wave height, H, was used to provide a non-dimensional expression.

Figure 11 provides a summary plot of $\Delta P/H$ against H/L_o for the four external pressure gauge elevations -20, -10, 0 and +10cm.

The maximum $\Delta P/H$ for elevation +10 cm was approximately 0.2 and is considered negligible since the wave runup seldom exceeded elevation +10 cm during testing. Maximum values of $\Delta P/h$ of approximately 2 to 2.2 occurred at small values of H/L_o for the other pressure gauge elevations. It is interesting to observe that the external pressure heads were not significantly influenced by gauge elevation; that is the values of $\Delta P/H$ appear to be quite similar for all ranges of H/L_o at elevations -20, -10, 0 cm. This suggests a relatively even distribution of wave energy on the relatively flat slope of the natural profile breakwater. This is quite different from what was observed in the conventional breakwater tests and may provide evidence as to why this profile is formed naturally.

(b) Comparison of the Results of Tests with Natural Profile and Conventional Breakwaters

Figure 12 provides a comparison of the variation of differential

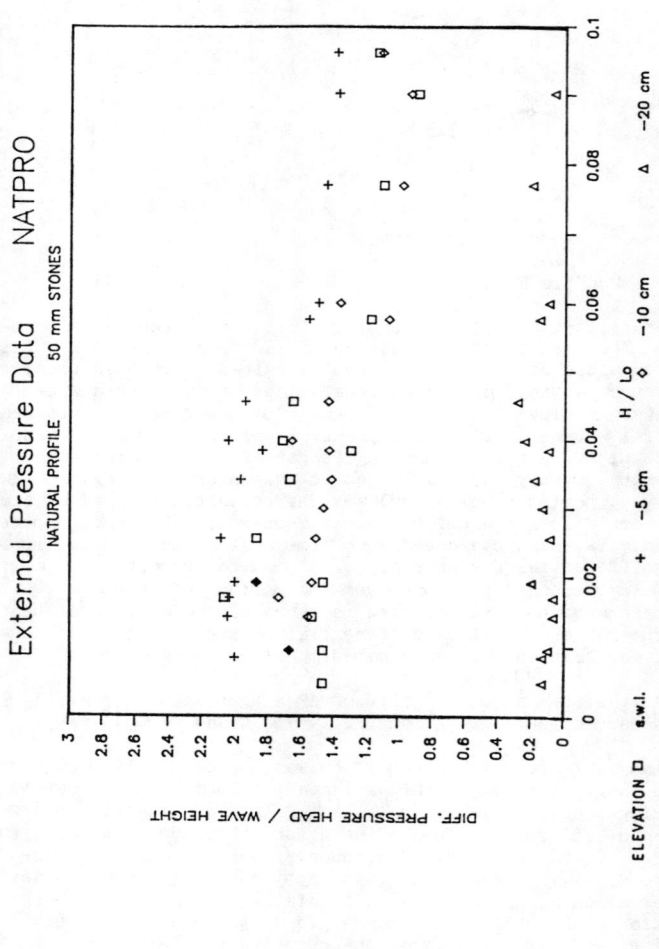

FIGURE 11 ΔP/H VERSUS H/Lo - NATURAL PROFILE TEST

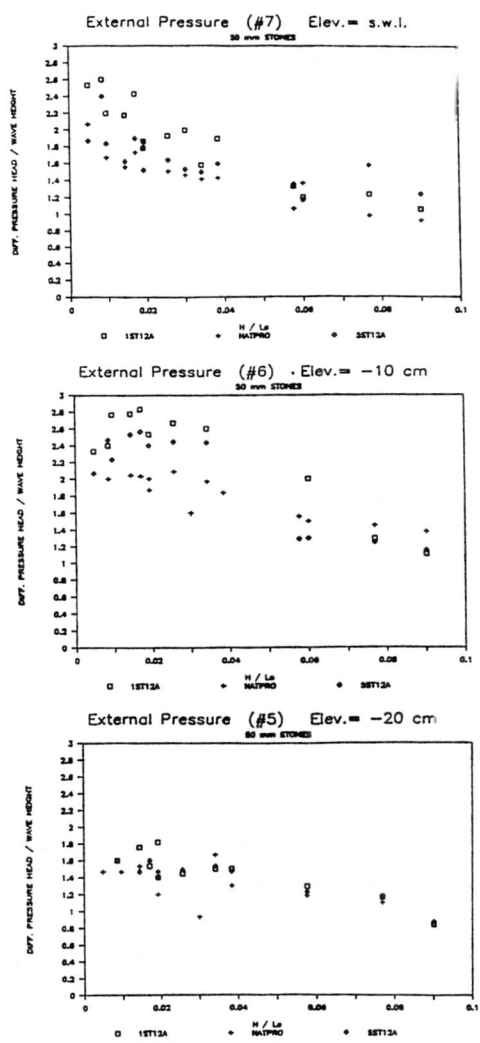

FIGURE 12 COMPARISON OF EXTERNAL PRESSURES MEASURED ON CONVENTIONAL AND NATURAL PROFILE BREAKWATERS

pressure head with wave steepness for the natural profile breakwater and a conventional structure constructed on a slope of 1:2. The data for the conventional structures is described in Hall, 1987. Figure 12 compares the data for the natural profile structure with the results of tests 1ST12A and 5ST12A (50 mm stones, 1:2 slope, 1 and 5 layers). This figure illustrates that for the given wave conditions and gauge elevation, the pressures measured on the natural profile are generally less than those measured on the conventional profile.

At elevation -20 cm, the variation of $\Delta P/H$ with H/L_o is fairly similar for the natural profile and the conventional breakwaters. At the still water level the values of $\Delta P/H$ for the natural profile test are consistently lower (15 percent on average) than $\Delta P/H$ values for test 1ST12A.

By comparing the results of the natural profile test with the results of the conventional structure tests, it was shown that values of $\Delta P/H$ for the natural profile test were less in the critical area of the structure (namely in the area of the still water level).

(2) Internal Pressure Measurements

(a) Natural Profile Results

The internal pressure acting along the core-filter interface of the test sections was measured at elevations -20, -10, 5, 0 and +10 cm. The internal pressure data was reduced using the procedures described earlier, so that the temporal pressure record for each test segment (wave height - wave period combination) was characterised by a single differential pressure head.

Figure 13 shows the variation of the dimensionless differential pressure head ($\Delta P/H$) with wave steepness (H/L_o) for each gauge elevation. These figures illustrate the same trends observed in tests of conventional structures, namely that $\Delta P/H$ decreased with increasing H/L_o. Generally, the rate of decrease of $\Delta P/H$ with H/L_o was less at higher values of H/L_o.

As was the case with the external pressure field, the variation of $\Delta P/H$ with elevation, for a given H/L_o value, was small; the exception being for elevation -20 cm where $\Delta P/H$ values were 30 to 40 percent higher than at other elevations. The relationship between the pressure response curves measured at all gauge elevations was examined. The pressure response curves measured at elevations 0, -5 and -10 cm were virtually identical whereas the differential pressure heads measured at elevation -20 cm were approximately 30-75 percent higher than those at the other gauge elevations, depending upon the wave conditions.

At H/L_o values of approximately 0.04, a large variation in $\Delta P/H$ values at elevation -20 cm was observed. The range of $\Delta P/H$ was 0.8 to 1.8. This may have resulted from the manner in which the wave interacted with the structure for the particular wave conditions (height and period) which occurred at this H/L_o. The pressure gauge was located on the steep slope section of the structure, 4 cm below the point of

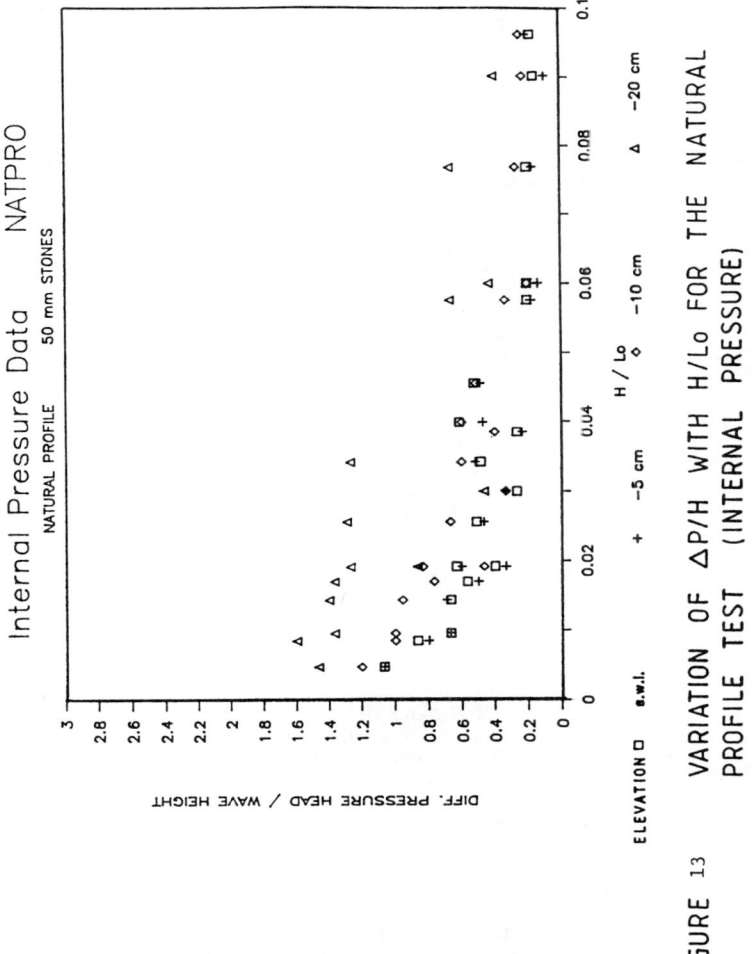

FIGURE 13   VARIATION OF $\Delta P/H$ WITH H/L_0 FOR THE NATURAL PROFILE TEST (INTERNAL PRESSURE)

the slope change. This localised geometric change may influence the flow kinematics resulting from wave breaking for these wave conditions.

(b) Comparison of the Results of Tests with Natural Profile and Conventional Breakwaters

The internal pressure measurements recorded on the natural profile structure can be compared directly with the results of series ST12A since the slope of the core (1:2), the armour type (50 mm stone) and the core material (type A - D_{50}= 3.5 mm) are identical. The internal pressure gauges were located at the same elevations in both test series. The relationship between the two structures is illustrated in Figure 8. The only subjective decision to be made was how many layers of armour on the conventional structure is representative of the natural profile structure. The problem arises since the natural profile has a variation in the number of layers of armour with elevation. Typically the armour layer varies from two stones thick at elevation +10 cm to 5 stones thick at -10 cm to 7-8 stones thick at -20 cm. For the purpose of comparing results, tests undertaken in series ST12A with armour layer thicknesses of 1 and 5 stones (the minimum and maximum cases) were used.

Figure 14 provides a comparison of the variation of $\Delta P/H$ with H/L_o for both the conventional and natural profile test sections (for each pressure gauge elevation). At all gauge elevations, the internal differential pressure heads measured in the natural profile test were consistently lower than the internal pressures measured in the conventional breakwater tests with both 1 and 5 layers of armour.

At elevation -10 cm for both the natural profile test and test 1ST12A and at all other elevations for test 1ST12A, localised increases in $\Delta P/H$ occurred at H/L_o values of 0.077 and 0.09. These increases may have resulted from the location of wave breaking or because of the manner in which the wave shape changed as a result of wave breaking. The actual wave conditions for these test segments were H = 90 mm and T = 0.8 sec for H/L_o = 0.09 and H = 120 mm and T = 1.0 sec for H/L_o = 0.077. Both of these cases were characterised by high wave heights and low wave periods and a resulting wave form approaching its limiting steepness.

(3) Phreatic Surface Movement

Figure 15 provides a comparison of the phreatic surface profiles measured during tests of the natural profile breakwater section and conventional type breakwaters. The behaviour of the natural profile structure is compared with that of the structures used in tests 1ST12A and 5ST12A for reasons given earlier. The phreatic surfaces measured during the natural profile test fell between the phreatic surfaces measured in the conventional designs with one or five layers of 50 mm armour stones. This was expected since the armour layer of the natural profile structure had a thickness of two to five stones depending upon the elevation.

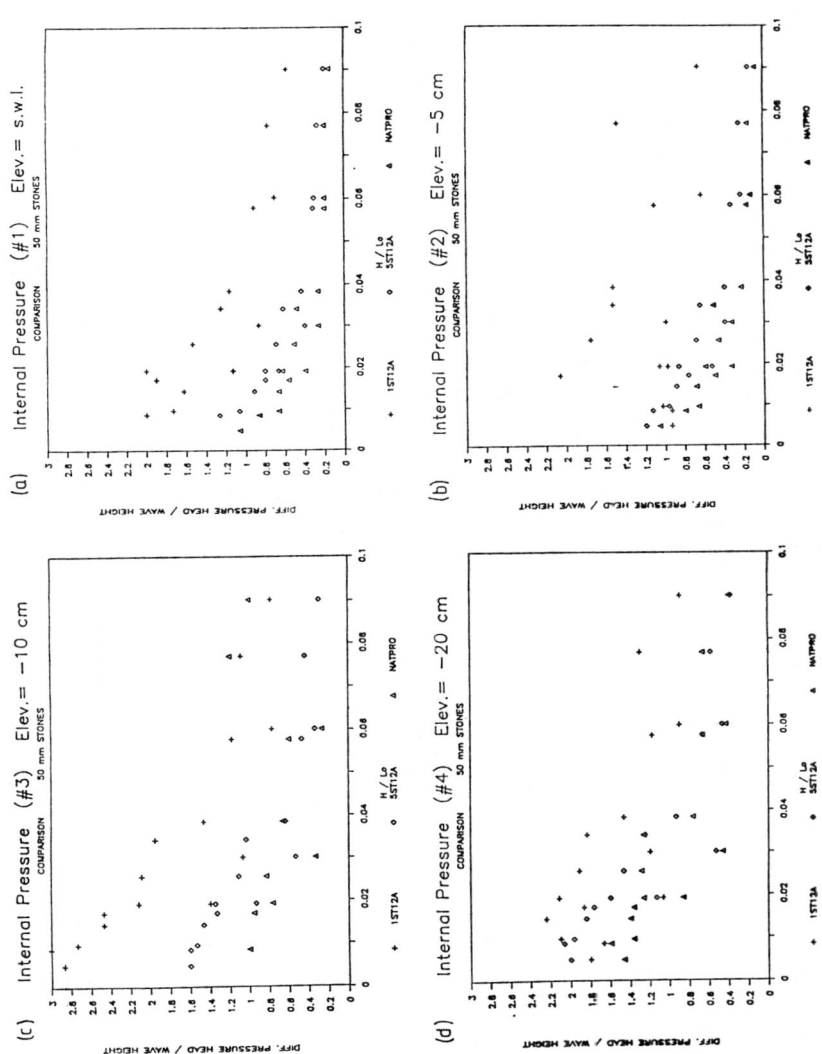

FIGURE 14 COMPARISON OF $\Delta P / H$ FOR NATURAL PROFILE AND CONVENTIONAL STRUCTURES

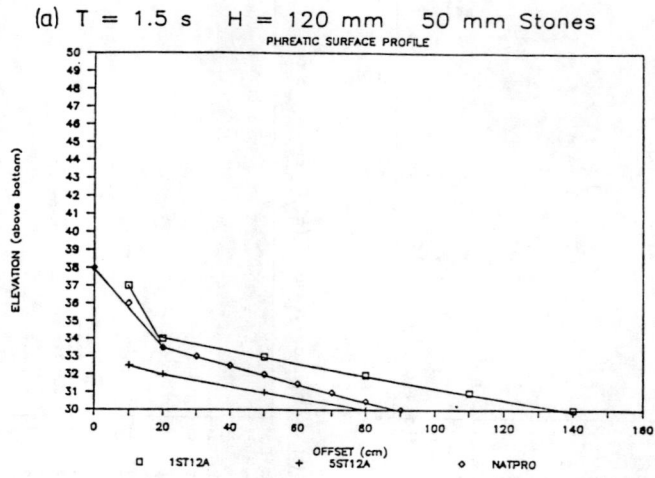

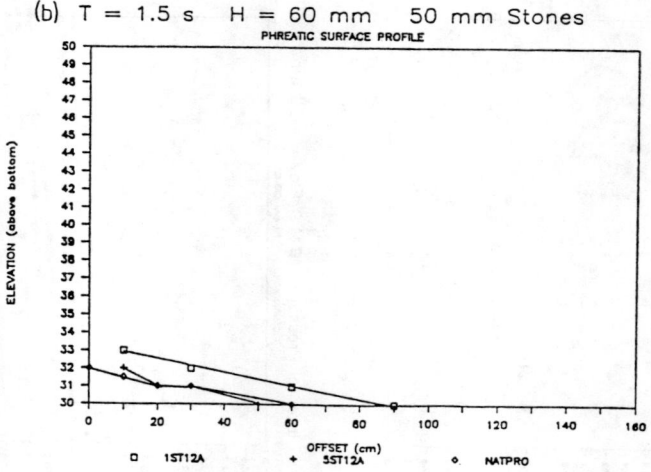

FIGURE 15 COMPARISON OF PHREATIC SURFACE PROFILES FOR CONVENTIONAL AND NATURAL PROFILE BREAKWATERS

VERIFICATION OF PERFORMANCE

Conclusions

The concept of a naturally armouring breakwater was introduced earlier. The relative advantages of this type of structure were described. Tests were undertaken to determine possible mechanisms from which this type of structure derives its stability. As a result of these tests, the following conclusions were made:

(1) Maximum values of the dimensionless external differential pressure head, $\Delta P/H$, of 2 to 2.2 were measured at small values of H/L_o. The variation of $\Delta P/H$ with elevation on the slope was found to be small; that is, the pressure at all gauge elevations was relatively constant. This is quite different from what was observed in the conventional breakwater tests, where maximum pressures were found to occur in the area of the still water level.

(2) At the still water level, values of external differential pressure head measured on the natural profile structure were typically 15-20 per cent lower than those measured on the outer slope of a conventional structure. This was shown to be the critical region of a conventional structure with respect to stability. The reduction of pressure in this area of the natural profile breakwater shows how profile readjustment, resulting from wave exposure, provides benefits in terms of overall stability.

(3) Values of internal pressure measured within the natural profile breakwater exhibited little variation with gauge elevation; the exception being at elevation -20 cm, where $\Delta P/H$ values were 30 to 40 percent higher than those measured at the other gauge elevations.

(4) For given test conditions, internal pressures measured in the natural profile breakwater were significantly less than those measured in a conventional structure. For H/L_o values less than 0.04, a reduction in $\Delta P/H$ of up to 50 percent was observed.

(5) The phreatic surfaces measured during tests on the natural profile breakwater fell between phreatic surface positions measured in conventional designs (1:2 slope) containing 1 and 5 layers of stone armour. This would be expected since the armour layer of the natural profile structure had a thickness of between two and five stones depending upon the elevation.

(6) The wave height at which the initiation of armour stone movement occurred was 60 percent higher in the natural profile test than that for a conventional structure armoured with two layers of the same sized stones. This specifically demonstrates that it is possible to use smaller stones on the flatter slopes of a natural profile structure than is possible on a conventional structure.

References

Ahrens, J.P., "Large wave tank tests of riprap stability", CERC Tech. Mem 51, May 1975.

Ahrens, J.P., "Reef-type breakwaters", 19th Int. Conf. on Coastal Eng., Houston 1984, pp. 2648-2662.

Baird, W.F. and Hall, K.R., "The design of breakwaters using quarried stones," 19th Int. Conf. on Coastal Eng., Houston 1984, pp. 1024-1031.

Bruun, P. and Johannesson, P., "Parameters affecting the stability of rubblemounds," Norwegian Inst. of Tech., Ports and Ocean Engineering Division, Report #1, 1974.

Foster, D.N., "Investigation of quarry run breakwater, Little Grassy, King Island, Univ. of New South Wales Water Research Lab Rept. 69/20, 1969, 17 p.

Gourlay, M.R., "Beaches: profiles, processes and permeability", Dept. of Civil Eng., Univ. of Queensland, Australia, 1980.

Hall, K.R., Rauw, C.I. and Baird, W.F., "Development of a wave protection scheme for a proposed offshore runway extension Unalaska, Alaska", Coastal Structures 83, Washington, March 1983, pp. 157-170.

Hall, K.R., "A Study of the Stability of Rubblemound Breakwaters", Ph.D. dissertation, Univ. of New South Wales, Sydney, Australia, May 1987.

Hall, K.R. and Foster, D.N. "Mass Armoured Breakwaters", Proc. 2nd Int. Conf. on Ports and Harbour Engineering in Sept. Developing Countries, Beijing, China, Sept. 1987.

Hijum, E. Van and Pilarczyk, K.W. "Gravel beaches: equilibrium profile and longshore transport of coarse materials under regular and irregular wave attack," Delft Hyd. Lab Pub. No. 274, July 1982.

Kemp, P.H., "The relationship between wave action and beach profile characteristics", Proc 7th Int. Conf. on Coastal Engineering, 1960 pp. 262-278.

Kogami, Y., "Researches on stability of rubblemound breakwaters", Coastal Eng. in Japan, Vol. 21, Dec. 1978, pp. 75-93.

McCorquodale, J.A., and Hannourra, A.A., "Virtual Mass of coarse granular media", ASCE Journal of Waterway, Port, Coastal and Oceans Division, Vol. 104 No. WW2, May 1978, pp. 191-200.

Moutzouris, C., "A profile of a sloping breakwater based on recent results concerning wave propagation and breaking," 7th Int. Harbour Congress, Antwerp 1978, pp. 2.04/1-2.04/7.

Naheer, E. and Buslov, V., "On rubblemound breakwaters of composite slope", Coastal Engineering, Vol. 7, 1983, pp. 253-270.

Popov, I.J., "Experimental research on formation by waves of stable profiles of upstream faces of earth dams and reservoir shores", 7th Int. Conf. on Coastal Eng., The Hague, 1960, pp. 282-293.

Priest, M.S., Pugh, J.W. and Singh, R. "Seaward Profile for rubblemound breakwaters", 9th Int. Conf. on Coastal Eng., Lisbon, 1964, pp. 553-559.

Rennie, J., "An historical, practical and theoretical account of the breakwater at Plymouth Sound," Weale, London, 1848.

Watts, G.M., "Laboratory study of effects of varying wave periods on beach profiles, Tech. Mem. 53, U.S. Army Waterways Experiment Station, Sept. 1964.

THE DEVELOPMENT OF A DESIGN
FOR A BREAKWATER AT
KEFLAVIK, ICELAND

by

W. F. Baird and K. Woodrow

Abstract

A breakwater design was developed to be built using local quarried rock and relatively simple construction methods. The design significant wave height was 5.8 m and the maximum depth of water was 24 m. The final design consists of a wide layer of 1.7 to 7 tonne armour stones in place of a traditional two layers or armour stones. At this location a traditional design would have required two layers of 30 to 40 tonne stones. The core of the breakwater contains the remainder of the quarry yield, that is, stone weighing less than 1.7 tonnes. The development of the design was supported by an extensive series of physical model tests.

The study demonstrates that by varying the geometry of a breakwater cross-section a design can be prepared that makes full use of the yield of a quarry and may use armour stones weighing five times less than the stones required for a conventional design. At this location this design approach achieved cost savings, compared to that for a conventional design, in the order of 40 per cent.

Résumé

Un brise-lames a été conçu de manière à pouvoir être construit avec de la roche d'une carrière locale par des méthodes relativement simples de construction. La hauteur significative nominale des vagues était de 5,8 m et la profondeur maximale de l'eau de 24 m. La conception finale retenue consiste en une large couche de blocs de pierre de carapace de 1,7 à 7 tonnes au lieu des habituelles deux couches de blocs de carapace. À cet emplacement la construction d'un brise-lames classique aurait exigé deux couches de blocs de 30 à 40 tonnes. Le noyau du brise-lames renferme le reste de la production de la carrière, soit les pierres de moins de 1,7 tonne. La conception de ce brise-lames était basée sur un important ensemble d'essais sur modèle physique.

L'étude démontre qu'en faisant varier la géométrie d'une coupe transversale du brise-lames, it est possible de concevoir un ouvrage construit avec l'ensemble de la production d'une carrière et utilisant des blocs de carapace d'une masse de 5 fois inférieure à celle des blocs utilisés pour un brise-lames classique. À cet emplacement cette méthode de conception a permis des économies de l'ordre de 40 pour cent comparativement au coût d'un brise-lames classique.

THE DEVELOPMENT OF A DESIGN FOR A BREAKWATER AT KEFLAVIK, ICELAND

W.F. Baird, P.Eng. (1) and K. Woodrow, P.E. (2)

(1) W.F. Baird & Associates Coastal Engineers Ltd., Ottawa, Canada.

(2) Bernard Johnson Inc., Bethesda, Maryland, USA.

ABSTRACT

A breakwater design was developed to be built using local quarried rock and relatively simple construction methods. The design significant wave height was 5.8 m and the maximum depth of water was 24 m. The final design consists of a wide layer of 1.7 to 7 tonne armour stones in place of a traditional two layers of armour stones. At this location a traditional design would have required two layers of 30 to 40 tonne stones. The core of the breakwater contains the remainder of the quarry yield, that is, stone weighing less than 1.7 tonnes. The development of the design was supported by an extensive series of physical model tests.

The study demonstrates that by varying the geometry of a breakwater cross-section a design can be prepared that makes full use of the yield of a quarry and may use armour stones weighing five times less than the stones required for a conventional design. At this location this design approach achieved cost savings, compared to that for a conventional design, in the order of 40 per cent.

INTRODUCTION

In 1981 the U.S. Navy, as part of an on-going design contract with Bernard Johnson Inc. required a protected harbor at Keflavik air base, Iceland. The harbor facilities required a breakwater to provide protection from wave action. In 1983, Bernard Johnson Inc. retained W.F. Baird & Associates to assist in the development of the design for the breakwater.

Helguvik Bay is located on the east side of the Reykjanes peninsula and is exposed to waves generated within the Flaxafloi, Figure 1.

The maximum depth of water at the location of the breakwater is 24 m, Figure 2. The tidal range is approximately 4 m.

The design wave conditions were determined from a wind-wave hindcast study based on winds recorded at Reykjavik and Keflavik airports, adjusted to represent overwater wind speeds. The breakwater was designed to successfully survive a storm with a return period of 50 years. The design wave conditions were determined to be as follows:

North east (waves attack at 45° to the breakwater centre line)
 $Hs = 5.8$ m, $Tp = 9.6$ s
East (waves attack perpendicular to the breakwater centre line)
 $Hs = 4.3$ m, $Tp = 7.2$ s

INITIAL DESIGN

A conventional design, based on the recommendations of the U.S. Corps of Engineers Shore Protection Manual (1984 edition) would require two layers of armour stones weighing 32 tonnes (slope of 1:2). Although this edition of the Manual had not been published when the initial design was developed and the design wave conditions were considered at that time to be lower

BREAKWATER AT KEFLAVIK, ICELAND

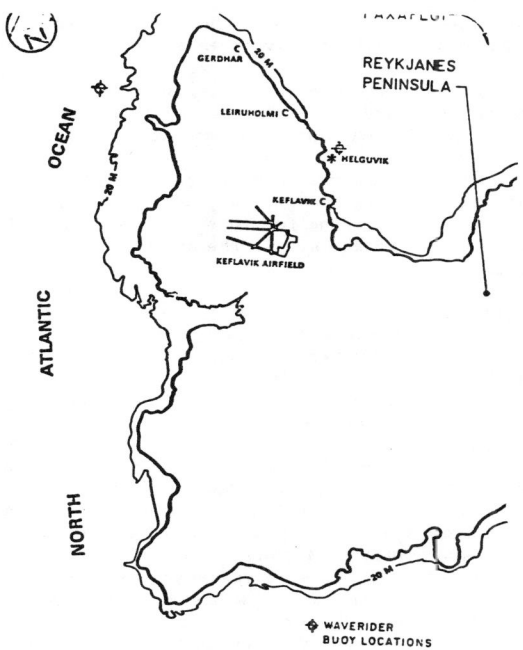

FIGURE 1   SITE LOCATION MAP

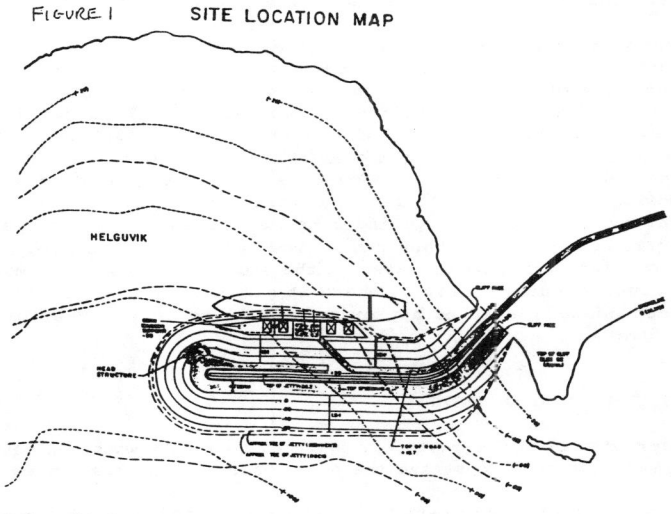

FIGURE 2

than those noted above, a conventional design would have required large armour stones that would not have been available from a local quarry.

Consequently, the cost of a breakwater would have been very high, because of the cost of transporting armour stone from a distant quarry (if one was found) or the cost of manufacturing concrete armour units.

WFBA had previous experience in developing breakwater designs that required relatively small armour stones, Hall et al, 1983, and an investigation was started to develop a design for this location using the yield of a quarry to be opened close to the site. The assumed yield of the quarry is shown in Figure 3.

This alternate design using smaller armour stones is based on the observation that if the armour layer is built to significantly greater thickness than that of two stones, much smaller stones are required to provide stable protection against wave action. Therefore, the thickness of the armour layer for a specific breakwater is determined based on the gradation of the available armour stones and the incident wave climate.

The relatively high porosity of the mass of armour stones allows the waves to propagate amongst the stones and dissipate their energy over a large area within the wide armour layer. In a conventional two stone armour layer, the flow produced by the incident wave is restricted by the relatively impermeable filter and core and, consequently, there are large velocities produced by the wave uprushing or downrushing within the narrow armour layer. In the berm the flow has a larger area into which it can move and as a result localized velocities are greatly reduced thereby decreasing the external hydrodynamic forces applied to the stones. A considerable increase in stability is achieved as a consequence of this dissipation of wave energy within the permeable mass of armour stones.

The mass of armour stones also increases its stability as a result of wave action. Wave action causes consolidation and a resulting increase in shear strength of the mass of stones. Motion of some stones at the surface results in "nesting" of the surface stones. This nesting process also results in an increase in the frictional restraint on individual stones. Depending on the size of stones available and the design wave conditions, movement of stones on the outer surface may occur to varying degrees. Movement takes place during the early stages of exposure to wave action. The stones eventually find a geometrically similar space in the berm surface into which they nest. The result of this process is a natural armouring of the outer layer of the stones. A typical armoured profile is illustrated in Figure 4, where it can be seen that the final profile has been consolidated to approximately 85 to 90 per cent of the as-placed volume.

PROGRAM OF MODEL TESTS

A program of model tests was designed to assist with the development of the design and to demonstrate the performance of the breakwater.

The tests were completed at a geometric scale of 1:35 in an 8 m wide wave flume at the National Research Council of Canada. Waves occurring from the

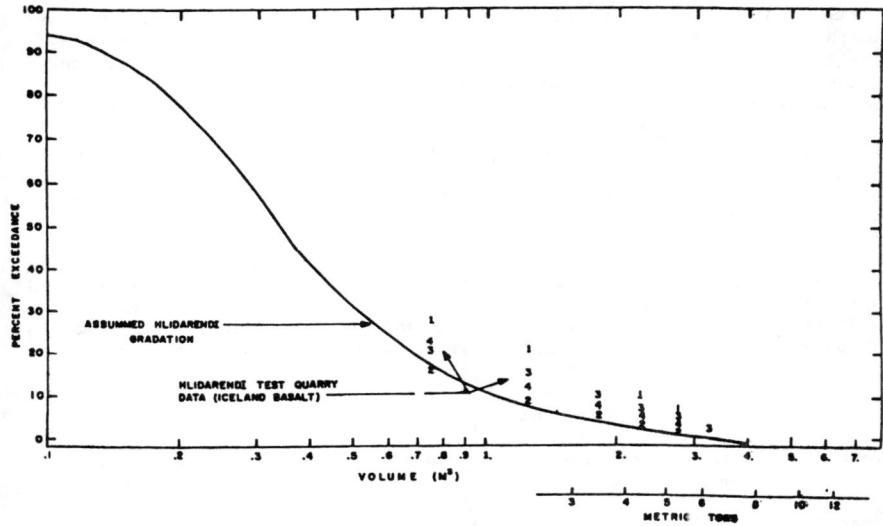

FIGURE 3   ESTIMATED QUARRY YIELD FOR THE HELGUVIK BREAKWATER

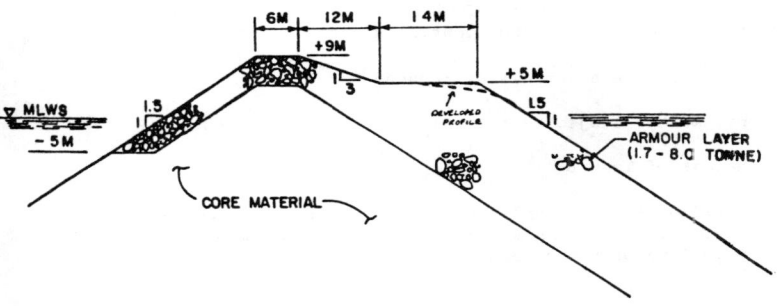

FIGURE 4.   "BERM" CROSS-SECTION

north east and east were generated. For the tests with waves from the north east the bottom topography in front of the breakwater was modelled. The full profile of the design storm, having a total duration of 40 hours, was simulated by nine segments in each test and irregular waves used for each segment of the storm.

The full breakwater, including the head of the breakwater was built in the wave flume. The armour layer was an accurate representation of the assumed prototype gradation. Each stone in the armour was individually weighed and sorted to provide the required gradation.

A total of ten complete tests with four additional demonstrations were completed. The tests were designed to evaluate the overall stability and performance of the structure when built using different stone gradations and subjected to various wave conditions and water levels. The information obtained as the tests proceeded were used to change and revise the breakwater design.

The ten tests are summarized below:

Test 1   Initial design with two gradations of armour stone subjected to the design storm from north east at water level of +4 m.

Test 2   Initial geometry retained and built with armour stone of 1.7 to 7 tonnes throughout.

Test 3   Model breakwater tested in Test 2 subjected to a) a second design storm, b) 24 consecutive hours of H_s = 5.8 m, T_p = 9.6 s (maximum wave conditions) and c) 14 hours of H_s >7 m and T_p = 9.6 s.

Test 4   Crest elevation reduced, water level increased to +5 m.

Test 5   Test of breakwater at low water level, i.e. +0 m.

Test 6   Test to evaluate the effect of localized placement of armour stones not meeting the gradation. One area contained 1.7 to 3 tonne stones, the other contained 6 to 7.5 tonne stones.

Test 7   Width of the armour layer reduced.

Test 8   Breakwater subjected to design storm from the east.

Test 9   Repeat of Test 8.

Test 10 Breakwater with reduced armour stone sizes to evaluate the influence of the armour stone gradation on the overall stability of the breakwater.

Quarry Test

A blast test program to establish quarry gradation curves was undertaken concurrent with the model test program. Ideally, the quarry tests would have been completed prior to the model tests; however, design schedule would not permit this for this project. Another local Icelandic basalt quarry gradation curve, considered to represent the local quarry was used as the initial model basis, the final blast test results generally confirmed the assumed gradation. A minor variation between the design armor stone required and the final quarry test gradation curve resulted in a design cross section adjustment as shown in Figure 4. This adjustment allowed for a projected 100% quarry utilization for the breakwater construction.

CONCLUSIONS

Following the completion of the model studies the following conclusions were drawn:

1. The breakwater will successfully survive repeated occurrences of storms with a return period of 50 years.

2. Tests were completed at water levels of 0, +4 m, and +5 m. The breakwater was stable for all water levels. Wave runup was predictably greater during the tests at high water. The maximum extent of wave runup during the high water tests was approximately +8 m to +9 m.

3. Tests were completed for two wave directions. The response of the breakwater was most critical for wave attack perpendicular to the centre line of the breakwater.

4. An armour stone gradation of 1.7 to 7 tonnes was recommended. Some reserve stability exists with this design and this was demonstrated using both lighter stones and reduced width of armour layer. This gradation was selected to allow full use of the quarry yield.

5. The head of the breakwater was carefully observed and its performance recorded in all tests. It was recommended that the full width of the horizontal berm of armour stones be placed around the head through 135°.

6. Localized placement of uniform sized stones, whether smaller or larger than the specified size did not affect the stability of the structure. It is anticipated that limited segregation of storms, as could occur during construction would not affect the performance of the structure.

It is useful to describe the performance of the breakwater during the design storm. No changes to the profile of the breakwater occurred during the first two segments of the storm ($Hs = 1.8$ m and $Hs = 3.05$ m). Some motion of stones was observed but was limited to "rocking-in-place". During the third segment ($Hs = 4.3$ m) some rounding of the outer edge of the horizontal berm occurred. This resulted from some stones initially placed in a relatively unstable position being rolled a distance down the slope during the downrush of the largest waves.

In the fourth segment ($Hs = 5.2$ m) more intense motion was observed. The stones that were removed were rolled to an elevation of 0 m to -10 m. However, as these stones were rolled out of position the net effect was to leave behind a more compact, nested and therefore stable outer layer of armour stones.

During the peak segment of the storm (Hs = 5.8 m), the observations were similar to the previous segment with the motion generally restricted to "rocking-in-place".

During the sixth segment (Hs = 5.2 m) and later segments the motion was considerably reduced compared to that observed with the same wave heights on the "up side" of the storm. When the design storm was repeated the structure was observed to be considerably more stable - in the sense that less motion was observed - than noted in the first storm.

The question of the durability of the stones was given consideration during the development of the design. Clearly the stones used in the model were considerably more durable than those used for construction. Principally this problem was addressed by testing the model with a gradation of reduced stone sizes as an approximation of an in-place gradation with broken stones. However, it was also observed that only a small volume of the placed armour stones actually moved or rocked and it is doubtful that the stability of the breakwater is dependent on the durability of these stones.

The validity of the physical modelling process was also questioned during the development of the design. Because of this concern all tests were completed with a large model (scale of 1:35) in a wide basin with irregular waves. Various Reynolds number criteria published in the literature, were satisfied and it was determined that the physical model would provide conservative results. This conclusion has subsequently been confirmed by undertaking additional tests with identical models at scales of 1:7 and 1:35.

THE DESIGN AND CONSTRUCTION OF A MASS ARMOURED BREAKWATER AT HAY POINT, AUSTRALIA

by

W. Bremner, Dr. B.A. Harper
and
Prof. D.N. Foster

Abstract

A mass armoured breakwater is defined as a rubble-mound structure that is designed and built in an initially unstable form, but with sufficient material provided to allow natural forces to modify its shape to a stable profile. During the whole process, the breakwater continues to perform its design function.

In this paper the design and construction of a prototype breakwater of this type is briefly described. The paper also explores the practicality of allowing the progressive interaction of design, physical model testing of design and construction to overcome the difficulties often encountered in the construction of rubble-mound structures when the rock source is undeveloped before construction is commenced. The results of the model testing, although somewhat limited to the solution of the specific problem, are further analysed to explore the relativity of high permeability in the design of structures of this type. This has resulted in the development of a computer model (HARBREM) which may be used as an initial design tool in the selection of armour sizes for physical testing. The development of this model is briefly described in the paper together with an example of model input and output.

Résumé

Un brise-lames à carapace en vrac est défini comme étant un ouvrage en enrochement conçu et construit suivant une forme initialement instable, mais avec suffisamment de matériaux pour permettre aux forces naturelles d'en modifier la forme jusqu'à l'obtention d'un profil stable. Pendant tout le processus le brise-lames continue de jouer le rôle pour lequel il a été conçu.

Dans cette étude, la conception et la construction d'un prototype de brise-lames de ce genre sont brièvement décrites. L'étude explore également les possibilités d'interaction progressive de la conception, des essais sur modèle physique et de la construction pour surmonter les difficultés souvent renontrées lors de la construction d'ouvrages en enrochements lorsque la source de roche n'est pas mise en valeur avant le début de la construction. Les résultats des essais sur modèle, quoique quelque peu limités à la solution du problème spécifique, sont davantage analysés afin d'explorer le caractère relatif d'une perméabilité élevée lors de la conception d'ouvrages de ce type. Cela a permis la mise au point d'un modèle informatique (HARBREM) qui peut être utilisé comme instrument initial de conception pour le choix des dimensions des blocs de carapace pour les essais physiques. La mise au point de ce modèle est brièvement décrite dans l'étude qui fournit également un exemple d'entrées et de sorties du modèle.

THE DESIGN AND CONSTRUCTION OF A MASS ARMOURED BREAKWATER AT HAY POINT, AUSTRALIA

ABSTRACT

A mass armoured breakwater is defined as a rubble-mound structure that is designed and built in an initially unstable form, but with sufficient material provided to allow natural forces to modify its shape to a stable profile. During the whole process, the breakwater continues to perform its design function.

In this paper the design and construction of a prototype breakwater of this type is briefly described. The paper also explores the practicality of allowing the progressive interaction of design, physical model testing of design and construction to overcome the difficulties often encountered in the construction of rubble-mound structures when the rock source is undeveloped before construction is commenced. The results of the model testing, although somewhat limited to the solution of the specific problem, are further analysed to explore the relativity of high permeability in the design of structures of this type. This has resulted in the development of a computer model (HARBREM) which may be used as an initial design tool in the selection of armour sizes for physical testing. The development of this model is briefly described in the paper together with an example of model input and output.

1.00 DESIGN ORIGIN

In cyclone "David" in January 1976, a conventional rubble mound breakwater was severely damaged at Rosslyn Bay. This damage was closely observed by the author of this design. Blain Bremner & Williams Pty Ltd (BBW) were commissioned to re-design this breakwater and supervise its construction. From experienced observation at Rosslyn Bay and subsequent model testing of the redesigned breakwater for Rosslyn Bay, it became clear why the damaged breakwater continued to protect the harbour. It was the first time (as far as is known) that a physical model was tuned to fail in the same way as the prototype (ICCE Hamburg 1978). Engineers have known for a long time that the empirical Hudson equation used in the design of breakwaters takes no account of wave period, wave grouping, wave direction, wave reflection and the real effects of permeability of the armour layers. The importance of wave grouping, in terms of damage to gravity type breakwaters, was first explored in a paper by J Ploeg (ICCE Hamburg 1978).

Consequently, designers of rubble mound type breakwaters rely heavily on their experience. Because of the influence of the Hudson equation and its use in the last twenty years or so, there is an increasing number of artificial units of complex and distorted shapes that are now commonly in use as armour units. These are propagating at the rate of a new unit every year or so. The design philosophy of these units is that because of their geometric shapes they have increased mechanical interlock forces as well as gravity forces giving higher stability factors for any given mass. Physical model tests frequently bear this out. It should be emphasised that the remarkable work by Hudson was never meant to be applied to armour units that rely for their stability on a large measure of interlock with one another. It is well known from the study of many breakwater failures around the world that the effects of mechanical interlock are very much influenced by wave period, wave grouping, wave direction and wave reflection. Failure modes of these units in armour layers are both unpredictable and frequently catastrophic. There is a body of experienced opinion that if suitable natural rock is not available, the only acceptable artificial armour material is in the form of concrete cubes or prisms placed in a random fashion so that the permeability of the armour layers is maximised. These armour units behave predictably under all conditions.

It seems that it is nature's way that artificial harbours more often than not are backed by low level coastal plains that are usually remote from good sources of natural rock. A notable recent exception to this is at Sines where good natural rock was adjacent to the site, but where artificial armour units were used and suffered catastrophic failure.

The use of any artificial concrete armour unit is usually very much more expensive than natural rock. The problem in Queensland and many other areas of the world for that matter, is that natural rock greater than 5 to 6t is difficult and costly to obtain. Natural rock of larger size usually has a very low yield from rock formations and from normal quarrying procedures.

This has been recognised for some years and the experience at Rosslyn Bay introduced the idea of using commonly available rock sizes, intermixed with modified concrete cubes with a grading that has the highest possible permeability, placed in a cross-sectional shape that eliminates or reduces the use of a crane and maximises the construction by end tipping the rock with a minimum amount of trimming by dozer and backhoe. The construction cross-section is designed so that natural wave action will reshape the seaward slope to the stable "S" shape found in nature.

This is also the basis of the design of the extension of the eastern breakwater at Townsville. This breakwater is designed to protect a reclaimed area including a container terminal, bulk gas terminal and similar port facilities. In this case, the breakwater is seawards of a revetted reclaimed area and separated from it by a stretch of open water. When the breakwater is reshaped to a stable configuration by nature, it becomes a partially submerged structure and as such, is capable of attenuating the incident wave height by a factor of at least 0.5. The revetment is designed to withstand forces due to this transmitted wave. This structure has a very high tolerance to waves very much greater than the design wave height.

2.00 TUG HARBOUR AT HALF TIDE

At present, ships mooring at the Hay Point Coal Loader are assisted by tugs which operate from Mackay which is nearly 20km away from the coal loader terminal. To improve the efficiency of ship berthing at the Hay Point Coal Loader it was decided to construct a tug harbour at the High Water Islet near the town of Half Tide. This will enable the tugs to operate from a shelter which is only 3km away from the coal loading facilities at Hay Point (Figure 1).

In the past, several design proposals for the construction of a tug harbour at Hay Point have been considered. In 1977 the Department of Harbours and Marine, Queensland proposed a tug harbour design consisting of two breakwaters each approximately 1km long. This design was model tested at the Water Research Laboratory, but the construction of the breakwater was not carried out. More recently a second design which consisted of a single breakwater approximately 350m long was proposed by the Department of Harbours and Marine, Queensland. This design consisted of a conventional breakwater having two layers of 12 tonne dolosse as primary armour on the seaward and leeward faces and 40 tonne concrete blocks on the crest. This design was model tested at the Department of Harbours and Marine's Laboratory at Deagon Queensland.

Preliminary investigations and trial blasts at a nearby quarry site at Mount Griffiths, which is 2.7 km from the harbour site, had shown that the maximum rock size available was of the order of 2 to 3 tonnes. Rocks of this size were then judged to be completely inadequate as primary armour in a conventional breakwater design.

Hence the Department of Harbours and Marine selected concrete dolosse and concrete cube units as the primary armour in the design.

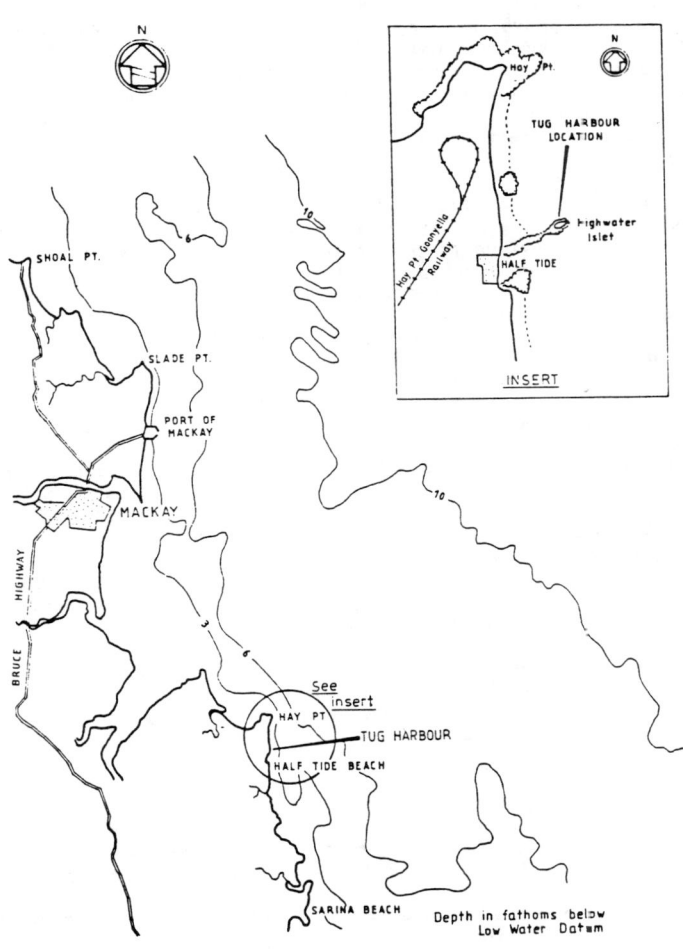

LOCALITY PLAN

FIGURE 1

However, following further investigations of the abandoned quarry at Mount Griffiths by a joint venture of the port users, Utah Development Company, Dalrymple Bay Coal Terminals Pty Ltd, in consultation with the Department of Harbours and Marine, the presence of more massive rock in andersitic dyke swarms was discovered. This led to the possibility of a redesign of a breakwater which could maximise the use of local material including large quantities of sound but heavily fractured rock mined in the process of extracting armour rock.

Blain Bremner & Williams Pty Ltd, (BBW) were engaged and developed a solution to meet this criterion which indicated a significant cost saving over all previous proposals and involved the design of a new breakwater section consisting mainly of mass armour stone. This is different from the conventional breakwater sections which normally uses only two layers of armour. It was also thought that by careful selection and blasting, the quarry may possibly produce some larger armour units. Hence the initial design of the mass armour breakwater section was based on rock units ranging between 3 to 7 tonnes.

The Half Tide breakwater completed in March 1987, embodies the collective experience of Grassy Island, BBW at Rosslyn Bay and Townsville and is designed on the basis that it will reshape under natural forces to a stable "S" shape. Physical model tests confirmed that this design results in a stable structure. This design also allows easy future maintenance of sections of the breakwater where reshaping by nature may be greater than in other sections. This usually occurs in local areas where there is a change of direction in the breakwater. This is often encountered in natural rock and shingle beaches for the same reason.

To test the performance of the proposed new mass armour breakwater and its stability under the design wave conditions, BBW requested Unisearch through the Water Research Laboratory to carry out hydraulic model studies to test the design.

From the earliest stages of models testing of Rosslyn Bay and Townsville breakwaters, the outstanding feature of this type of design was its extraordinary tolerance to wave forces very much higher than the design forces. It is this feature that has excited the interest of the Engineers who are the Authors of this paper.

The Half Tide breakwater during the model testing has been subjected to wave heights much greater than those depth-limited waves that can occur naturally on the site at Half Tide.

This breakwater design therefore, has the following basic characteristics:

1. It proposes the use of natural rock in commonly occurring quarried sizes.
2. It reaches stability by reshaping by natural forces.
3. It has a very high tolerance of forces very much larger than the design forces.

4. It is capable of very easy and relatively inexpensive maintenance by end tipping the rock. It also can be readily increased in height and length by the same method.
5. It results in large savings in capital cost against any other known design solution.

The variation in stability between a conventional two layered structure and a mass armour structure can be attributed mainly to the difference in the permeability of the two structures and the volume of material used. This results in:

i) Lower wave run up
ii) Lower wave reflection
iii) Increased absorption of wave energy within the body of the structure
iv) Higher transmission of wave energy to the leeward side
v) Reduction of breakwater slope following damage.

All these above factors were clearly evident during the testing of the mass armour breakwater.

It is well known that as the water level fluctuates on the face of the structure between run up and run down levels, a higher run up level causes steeper pressure gradients. This leads to higher seepage and draw down forces on the armour units. Further, for a given wave period, the velocity of water rushing down the face of the structure increases with increasing run up. As the drag force on the armour units is a function of the square of the fluid velocity, increasing run up will also increase the drag forces. Hence by reducing the run up, the stability of the armour units is improved by reducing both seepage and drag forces. Another influence not yet investigated is the influence of the compression of entrapped air in the interstices of the armour rock.

The wave height resulting at the structure is a combination of both the incident and the reflected waves. As the reflected wave height is reduced, the wave height acting on the structure is also reduced leading to a more stable structure.

The reflection from a conventional two layered breakwater with a similar sloping seaward face may is of the order of 40% whilst the reflection from a mass breakwater may be as low as 10%. Thus for a given incident wave height H, the wave height acting on the structure for a mass breakwater section is of the order of 1.1 H, whilst for a conventional breakwater, the wave height on the structure will be of the order of 1.4 H. If as stated by Hudson the stability of an armour unit varies as H^3, this influence alone may double the mass of the armour unit.

Unlike the conventional breakwaters where only two primary armour layers overlay a secondary layer and an impervious core, in a porous mass armour breakwater, energy can penetrate into the body of the structure. Hence the concentration of wave energy on the top layers of the breakwater is significantly reduced and the stability of the exposed armour layers is increased.

When viewed from the point of structure stability as against stability of individual armour units, a structure having many layers of armour units is capable of absorbing wave energy for a longer period when compared with a two layered structure which will be quickly destroyed once the two outer layers are removed and the filter and core becomes exposed. This was recognised by Hudson in the values given for K_D for 2, 3 and 4 layers of armour.

In a structure with an impervious core, wave energy is not transmitted through the body of the structure to the leeward face. The high permeability of the mass armour breakwater results in transmission of wave energy through the structure to the leeward side of the breakwater. For the breakwater sections tested during this investigation, the transmitted wave heights for incident wave heights of 3.6m at MSL and 6m at RL 4.5m AHD, transmitted wave heights were of the order of 0.1m and 0.6m respectively.

The transmission of wave energy, although resulting in some minor wave action on the leeward face, reduces the wave energy which otherwise would have been dissipated on the outer face. As the amount of energy which needs to be dissipated on the outer face of the structure is reduced, it will also reduce the forces acting on the exposed armour layers, thus increasing the stability of the structure.

The stability of the leeward face is not significantly affected by the transmitted wave as these wave heights are relatively small compared to the incident wave for which the whole structure is designed.

At wave heights in excess of that required to initiate damage, the seaward face tends to develop an "S" shape with relative flat slopes at the still water level (SWL) with steeper slopes both above and below. For a conventional breakwater, maximum damage occurs in the vicinity of SWL where wave forces are a maximum and it is the slope at this location which is the critical parameter in Hudson's equation. Reducing the breakwater slope in this area is known to increase breakwater stability and several investigators have suggested that for a conventional two layer breakwater increased stability can be achieved by using an "S" shaped seaward profile rather than one of constant slope. In the mass armour breakwater this shape develops naturally during reshaping by nature without damage to the core or secondary armour which would occur in a conventional two layer breakwater.

The influence of permeability on breakwater stability is presently poorly defined mainly because of the difficulty of modelling flow through porous media in a Froude model. For coarse materials where head losses in the prototype are proportional to velocity squared (i.e. independent of viscosity), a Froude model will give useful answers provided a sufficiently high Reynolds number is used.

At the scales chosen for the model studies, the scale effects are considered to be very small and are lower than model testing of a conventional breakwater, where it is impossible to model correctly the permeability in the core and secondary armour.

The main thrust of this design is its higher factor of safety against wave forces greatly in excess of the design waves and its ability to protect the area in its lee, even when severely damaged. Repairs and maintenance are readily and economically carried out. Every maintenance repair also increases the stability of this type of structure.

BBW were granted a research grant in 1985 by The Australian Marine Sciences and Technologies Grants Scheme to further research the Stability of Highly Permeable Breakwaters. Mr W Bremner and Dr B A Harper of BBW were the participants. The research work under this grant is also briefly outlined in Section 8.

3.00 NUMERICAL WAVE MODELLING

Although waverider recording systems had been operated in the area over a number of years, data on severe storm and tropical cyclone events was not extensive, with the highest recorded significant wave heights only of order 2.5m. However, even the relatively low intensity storms of decayed tropical cyclone Otto (990 mb) in March, 1977 and tropical cyclone Kerry (994 mb) in February-March 1979 generated waves of sufficient height to cause minor damage at the Utah Berths at nearby Hay Point. Because of the much higher waves likely to occur, numerical wave height prediction methods were used in determining suitable design wave parameters for the breakwater.

The wind wave predictions were performed using SPECT, a numerical spectral wave model originally devised by James Cook University (Young and Sobey) but also extensively developed by BBW.

For this project, a numerical model was formulated which extended from Cape Townshend north to the Whitsunday Islands and seawards to the Great Barrier Reef covering an area of over 50,000 square kilometres (Figure 2). Within this area model tropical cyclones were directed at Half Tide in an attempt to determine the highest possible wave heights for various tropical cyclone intensity, sizes and speeds of approach. Two basic types of storm were considered as shown in the model grid - Figure 2.

i) Classical coast-crossing tropical cyclones which move directly onshore. Four separate approach directions of N, NE, E and SE were used.
ii) Low intensity slow moving or offshore low pressure systems- such as decayed tropical cyclone Otto.

Storm intensity was based on previous research carried out by BBW, into the probability of occurrence of severe storm events along the Queensland coast.

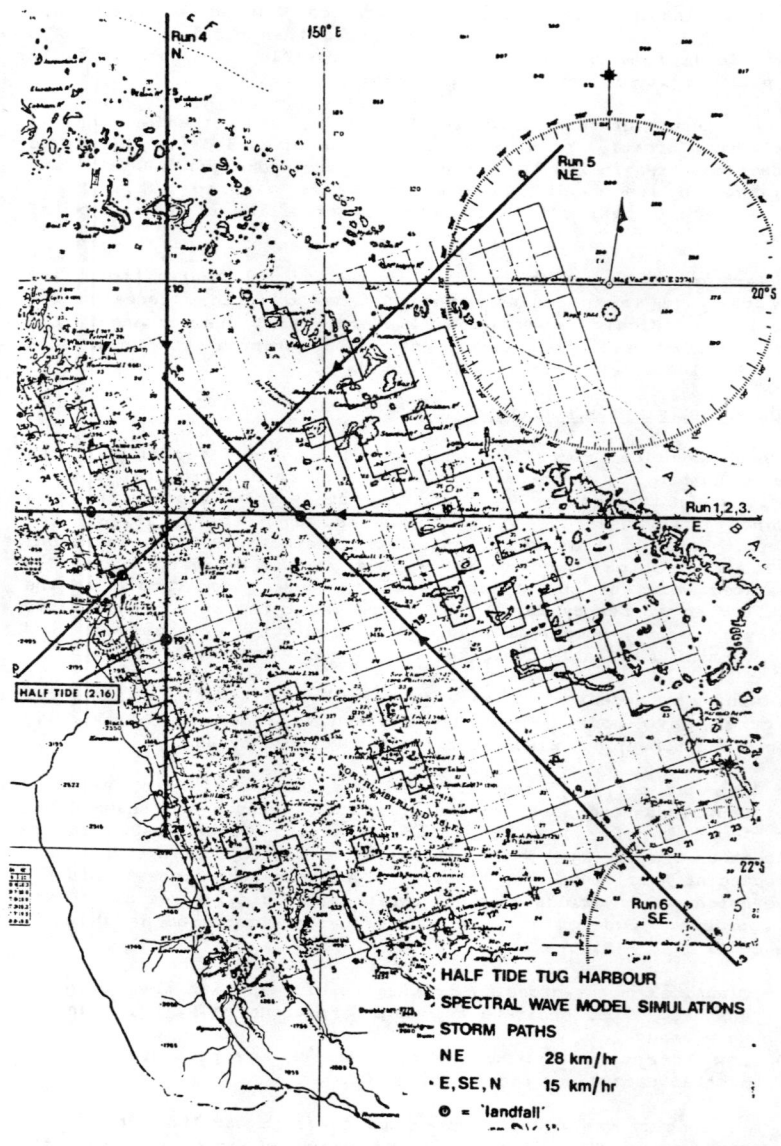

FIGURE 2.

Results for the Mackay region, which has the highest frequency of severe events, were representative at Half Tide and a 100 year design storm of 950 mb was determined for an area within 100 km of the breakwater site.

The east approach storm produced the highest significant wave at Half Tide for the 100 year storm of 4.7 m followed by 4.0 m for N, 3.9m for NE and 3.4 m for SE approaches. Very complex patterns of wave height contour and direction were forced by the near circular wind field of the storm. Together with the storm movement effect the areas of wind exposure constantly change throughout the simulation because of the numerous islands, shoals and the Great Barrier Reef. In most cases the area of highest waves was approximately due east of Mackay in the vicinity of Bailey Island.

The model highlighted the directional variation of the wave pattern but also showed energy shifts within the spectrum, as shown in Figure 3, from one wave length to another as the sea-state becomes more fully developed. In particular the NE approach indicated the influence of locally generated seas as the region of maximum winds neared the coast while the SE approach produced offshore wind for much of its approach with onshore waves only occurring after the storm passed to the north of Half Tide.

The worst approach (E) storm was re-run with a slightly higher mean water level of MSL +3m. This resulted in a higher significant wave at Half Tide of 5.0m as shown in Figure 4. The increased depth, being representative of a surge and/or tide combination, modified the wave paths approaching the shore and also reduced wave attenuation particularly in the extensive shoal areas SE of Half Tide.

The E approach for a 500 year return period storm of 925 mb resulted in a peak significant wave at Half Tide of 5.8m. This can be compared to a 10 year return period value of 990 mb which produced a 2.7m significant wave. For a stationary storm the direction of wave approach was similar to the E approach storm but the constant wind speed and direction produced a complex pattern of wave heights throughout the area. Of particular interest was the energy shift experienced at Half Tide resulting in a peaking of the significant wave height and then subsequent decrease while offshore waves continue to build. The observed variation of wave height with central pressure was an essentially linear response over the range of central pressures tested.

The results of the spectral wind-wave modelling in the vicinity of the proposed Half Tide Tug Harbour can be summarised as follows:-

1) Highest waves at Half Tide are indicated for storms approaching from the east and making landfall north of the site near Cape Hillsborough
2) The peak 100 year event at Half Tide indicates a peak significant wave of order 5m with peak period of order 7 sec and bearing 254°.

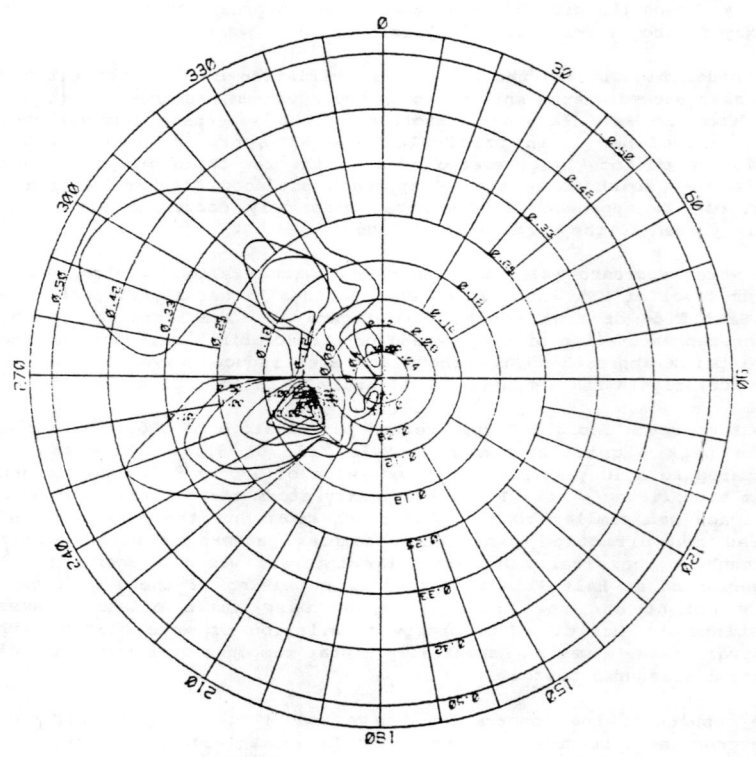

FIGURE 3

ARMOURED BREAKWATER

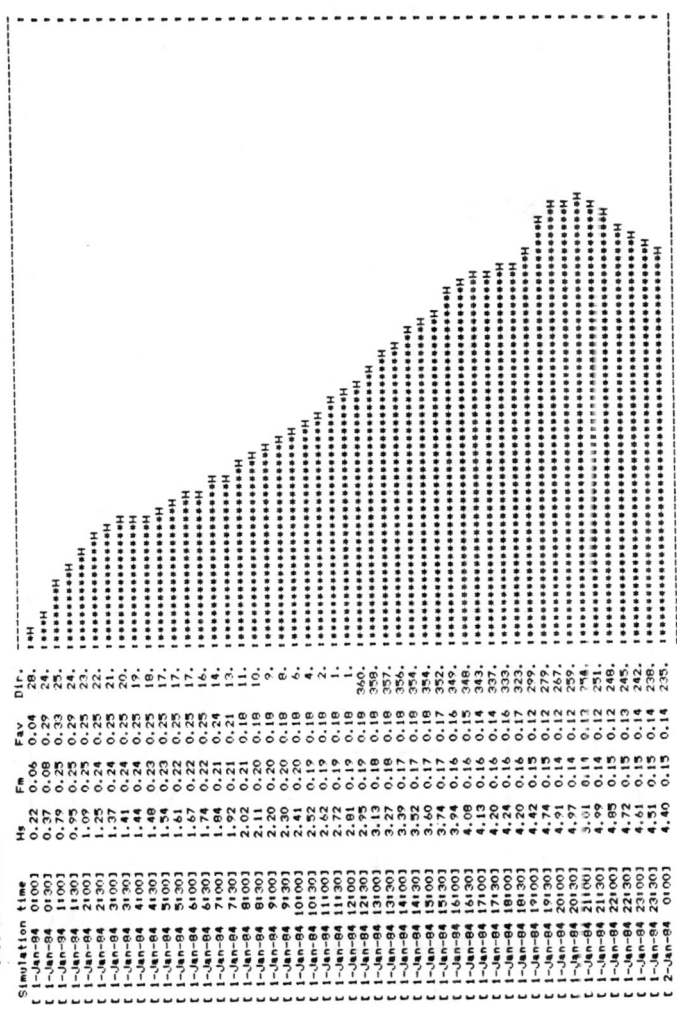

FIGURE 4.

BERM BREAKWATERS

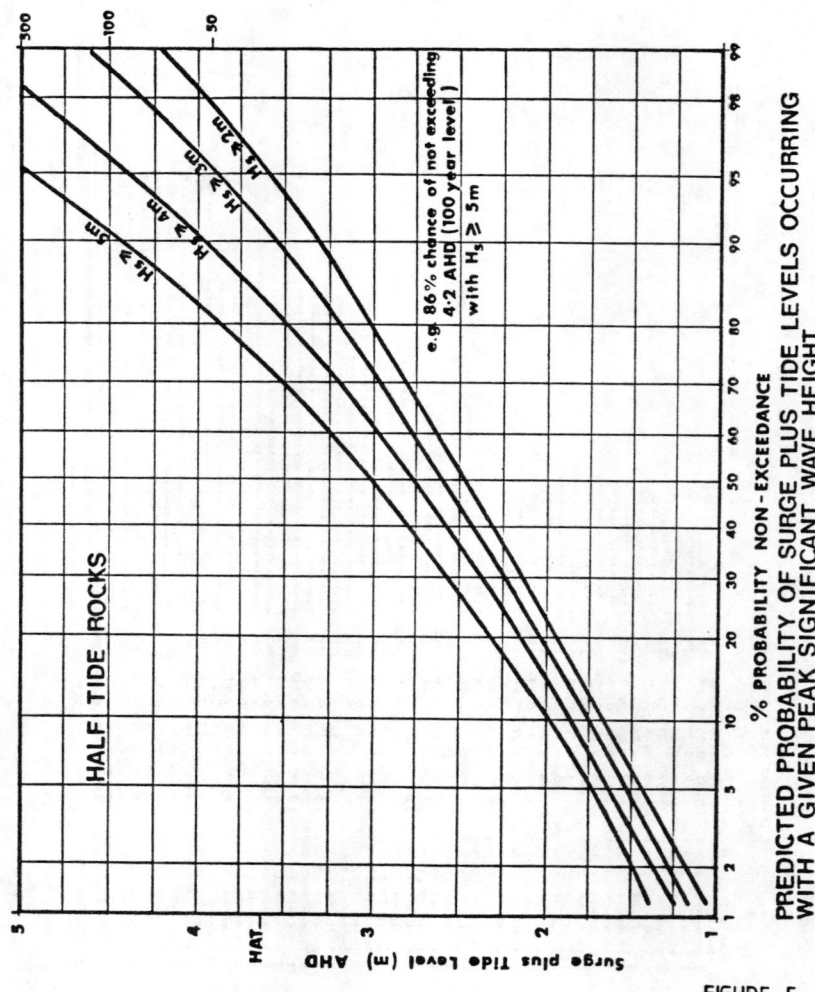

FIGURE 5.

3) In all simulation runs peak waves at Mackay are predicted as somewhat higher than Half Tide (order 1m) with longer peak periods (0.5 to 1.5 s). In particular, for the relatively rare SE approach storm, Half Tide is considerably more protected than Mackay. At various stages of storm approach, differences in significant wave height between these two sites of up to 2m are often experienced.

Storm surge levels at the breakwater site were also estimated using numerical modelling techniques in a two stage process:-

1) The result of Beach Protection Authority storm surge levels due to particular tropical cyclones were compiled to produce a set of surge response factors for the breakwater site in terms of storm speed, intensity, distance from site etc.
2) The BBW statistical simulation model SATSIM was then used to simulate the occurrence of tropical cyclones in the area and to estimate their resulting storm surge for a period of 15,000 years. This was achieved by projecting forward the historical record of storm parameters and randomly "generating" events which, over the long simulation period, closely resembled the statistical makeup of previous storms.

The 100 year event design water level for surge effects alone at Half Tide was determined to be 4.2m AHD.

The storm surge results were also combined with the predicted wave heights to assess the influence of wave setup levels at the breakwater site and also to determine the probability of waves of certain heights attacking the breakwater coincidentally with a storm surge, as shown in Figure 5.

4.00 HYDRAULIC MODEL TESTS

The following model tests were carried out by Unisearch through the University of New South Wales Water Research Laboratory at Manly Vale Sydney under the direction of Professor D N Foster. The authors initiated the designs to be tested and assisted in the supervision of the testing. Full reports of the tests are contained in the Technical Reports of the University nominated under.

Test Series 1. Technical Report No. 83/15 January, 1984
Test Series 2. Technical Report No. 86/02 May, 1986
Test Series 3. Technical Report No. 86/08 August, 1986.

MODEL TEST SERIES 1

Test series 1 was conducted using a design by BBW after an independent geological investigation of rock resource at Mt Griffiths indicated that the yield armour from 4t to 7t would be of the order of 10% to 15%. Estimated costs of this design indicated that a mass armour breakwater was significantly lower than that of previous designs using conventional two layer armour design and artificial armour units.

This design outline is shown in Figure 6 with a typical cross-section of the trunk of the breakwater in Figure 7.

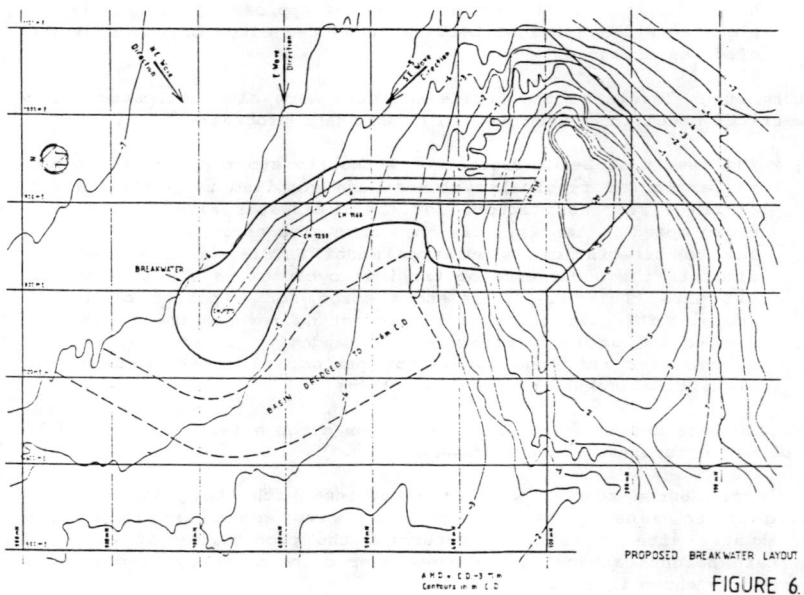

PROPOSED BREAKWATER LAYOUT

FIGURE 6.

Breakwater Geometry
─────────────────────

The breakwater is connected to the land at the High Water Islet and extends approximately 350m in a northerly direction towards the -6m Chart Datum (CD) depth contour. The headland at the High Water Islet shields the breakwater from the South Easterly waves.

Inside the harbour the bed is dredged to RL -6m CD (-9.11m AHD) to provide sufficient draft for the tug boats. Along the seaward face of the breakwater the water depth varies between 10 to 14m for the 1:100 year storm tide level of RL 4.5m AHD.

The breakwater section is made of a core consisting of quarry stone smaller than 2 tonnes and primary armour ranging between 3 to 7 tonnes. (Figure 7). The same material is used in the trunk and the head sections of the breakwater. The core is 16m wide at the top with side slopes of 1V:1.35H. The crest of the core is set at RL 0m AHD. The crest of the breakwater is set at RL 6.39m AHD and is 25.5m wide. In the trunk section of the breakwater the ocean face is sloped at 1V:1.35H and the leeward face is flatter at a slope of 1V:2H.

The head section is sloped at 1V:3H and has a semi-circular shape in plan form. In the transition section between the head and the main trunk section to the side slopes vary between 1:3 to 1:1.35 in the seaward face and 1:3 to 1:2 on the leeward face.

The design intention was to construct the core as shown in Figure 3 with two layers of armour on the outer face. This stage 1 construction was to be completed at the fastest practicable rate to give protection to dredging and harbour infrastructure and for this construction to commence at the earliest date.

Design Data

At the time of commissioning the hydraulic model study, BBW was undertaking a tropical cyclone wave modelling study (para 3.00) to arrive at the design wave conditions at the breakwater site.

As the cyclone wave modelling investigation was still continuing at the start of the hydraulic model study, the following preliminary design data was provided by BBW for testing the hydraulic model:-

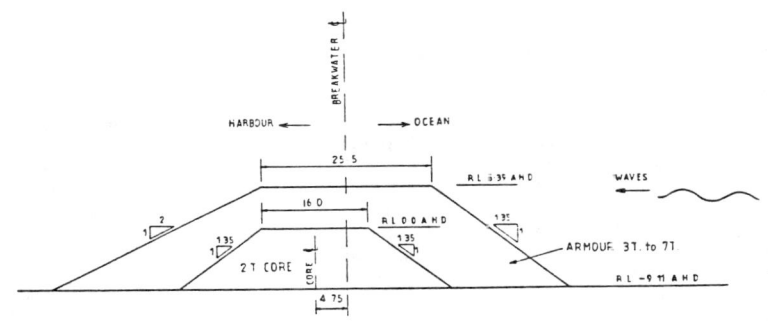

CROSS SECTIONAL GEOMETRY OF PROPOSED BREAKWATER

FIGURE 7

Design Wave Height H_s for 100 year return period = 6m
Wave Periods 8 and 11 sec
Wave Directions SE, E and NE
20 Year Return Storm Tide Level RL 3.75m AHD
100 Year Return Storm Tide Level RL 4.50m AHD
Tidal Range 6.5m

Initial model testing was undertaken using these design conditions.

After completion of the cyclone wave model study the following revised design data was provided and used in the tests of the final design:

Significant Wave Height = H_s = 5m
Peak Spectral Wave Period = 7 sec
Dominant Wave Direction N 74 E

Significant wave heights in excess of 7m are virtually unobtainable at the site.

During the model study both 3D wave basin tests of the complete breakwater as well as 2D wave flume tests on several breakwater sectional geometries were undertaken. In all 12 series of tests were completed.

All model tests except the final flume tests, were carried out using monochromatic waves. The water level corresponding to various surge levels was simulated by increasing the water level in the model by discrete steps.

Testing was carried out to investigate the stability of breakwater under various storm conditions for waves approaching from SE, E and NE directions. Figure 9 is a typical cross-section. For each wave direction two wave periods 8 sec and 11 sec were tested. It is possible for the prototype to experience several minor storms that can reshape the breakwater to an extent which may affect the stability when exposed to the design wave. Hence it was decided to simulate these conditions by testing the breakwater at lower wave height and storm tide levels before testing at the 100 year design storm condition. Since the stability tests were carried out using monochromatic waves, the higher waves in the group were simulated by testing the breakwater with 9m waves. For each wave direction and wave period the breakwater was tested for a period corresponding to a 7.5 hour storm.

For all wave and storm tide conditions tested the breakwater was stable and the damage was limited to the seaward side of the breakwater centre line. Even under the worst test conditions the damage did not extend beyond the crest centre line at any point along the length of the breakwater. The maximum damage occurred when the 6m wave was plunging on the structure.

ARMOURED BREAKWATER

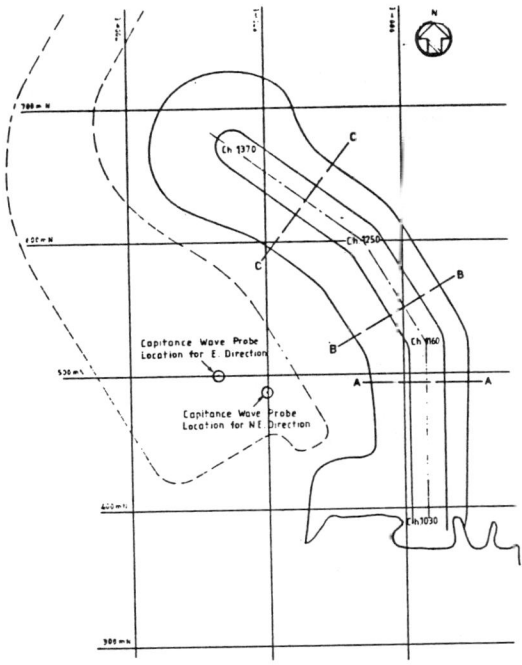

CROSS-SECTIONAL PROFILE LOCATIONS
AND LOCATIONS WHERE HARBOUR WAVES MEASURED

FIGURE 8.

The 8 sec waves from the easterly direction caused the most damage on the structure. At this wave period the higher waves were plunging directly on the structure and reshaped the seaward face of the structure between Chainages 1160 and 1250. The crest on the seaward side was cut back by nearly 8m.

The breakwater head section was extremely stable and there was no measurable change in the geometry for any of the wave directions tested. Similarly, there was no change in the leeward face or the leeward section of the crest.

The highest waves in the group were breaking before reaching the structure, even at 1:100 year storm tide level tested. Due to this depth limited conditions, any significant overtopping and/or damage due to the large waves in the group is unlikely.

Observations during the test revealed the mass armour breakwater to be an efficient dissipator of wave energy. Waves which travelled along the structure dissipated without becoming steeper and breaking on the structure and there was minimum reflection from waves which came directly onto the structure. Waves which ran on to the crest were rapidly absorbed without causing any significant overtopping.

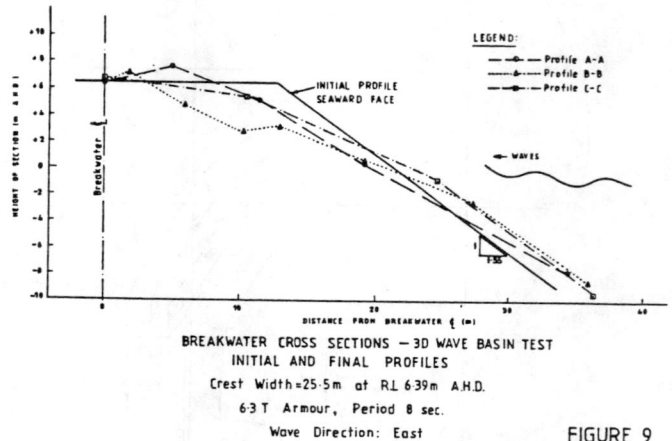

BREAKWATER CROSS SECTIONS — 3D WAVE BASIN TEST
INITIAL AND FINAL PROFILES
Crest Width = 25.5m at R.L 6.39m A.H.D.
6.3 T Armour, Period 8 sec.
Wave Direction: East

FIGURE 9

Due to the extremely high permeability of the structure there was only minimal overtopping even at the highest storm tide level of RL 4.5m.

Wave heights within the harbour were measured for 8 and 11 sec period waves from the NE and 7 sec period waves from the E.

Waves within the harbour were caused by wave diffraction around the head of the breakwater, wave transmission through the permeable breakwater and wave overtopping. As the offshore wave height increased the wave height due to each of the above factors also increased thus leading to a higher wave within the harbour. However, in the absence of any significant overtopping the main factor which caused waves within the harbour was wave diffraction. Wave diffraction is a function of the wave length which in turn is dependent on the wave period. The effect of wave diffraction inside the harbour was clearly evident from the higher waves which result at 11 sec period compared to the 8 sec period.

All 2D tests were carried out in a 1m wide laboratory flume. The test sections were built on the 1:100 sloping floor section of the flume which represents the approximate bed slope seaward of the structure. For these tests the bed level at the seaward toe of the structure represented approximately -9m AHD.

The wall level and wave height combinations for these tests are the same as those described for the 3D tests. The model material was the same size and grading as that used for the 3D tests. For the 6.3T armour test the linear scale was 60 which is the same linear scale as for the 3D test series. For the 5T armour tests, the linear scale was 55.7. For the 4T armour tests the linear scale was 51.6.

Under these conditions the maximum wave height which reached the structure was of the order of 7 to 8m. Even under the worst conditions there was only minor overtopping and some splashing on the leeward side. This did not cause any significant damage on the leeward section of the crest or the sloping face. There was small but measurable damage on the seaward face and the seaward section of the crest.

The damaged breakwater profiles at three sections were taken at the end of nearly 8 hours of testing. The results compared well with those from the 3D wave basin tests and confirms the greater stability and high resilience of the structure compared to that of a conventional two layered structure. The significant hydraulic characteristics, namely the lower run up, extremely high permeability and absorption and little or no wave reflection from the structure were also noted during these 2D tests.

The results from the 25.5 wide crest showed that only about 5m of the seaward crest width was damaged and that there was no significant damage on the leeward face. As the extent of damage was relatively small compared to the total area of the breakwater it was decided to undertake tests to optimise the sectional geometry of the breakwater. Hence tests were carried out on a 17m wide crest.

Except for the reduced crest width all other dimensions including crest elevation, armour size and model scale and test conditions are the same.

The damage in the seaward crest section for the same distance of 5m and this still left nearly 12m of undamaged crest width.

Due to the narrower crest width there was slightly more overtopping and splashing on the leeward crest but this was not sufficient to cause any significant damage on this section of the breakwater. The increased splashing also resulted in slightly higher wave heights in the harbour side which reached to nearly 0.6m.

The final sectional profiles after 8 hours of testing are shown in Figures 9 and 10. These profiles show a marginally greater damage than for the equivalent 6.3T armour section. This was partly due to the test being carried out for a longer duration with plunging wave conditions which caused the most damage. The seaward crest was cut back by 6.5m to reach within 2m of the crest centre line leaving an undamaged crest width of approximately 10m.

The reduced crest width also resulted in greater overtopping of the structure but this overtopping was still not sufficient to cause significant damage on the leeward section of the breakwater.

Except for marginally greater damage, the damaged profiles for the 5T and 6.3T armour breakwater sections were similar. The stability of the two armour sizes also did not appear to be significantly different. It is likely that in a breakwater consisting of mass armour the weight of the armour units plays a lesser role in the stability of the breakwater section compared with that of a conventional breakwater.

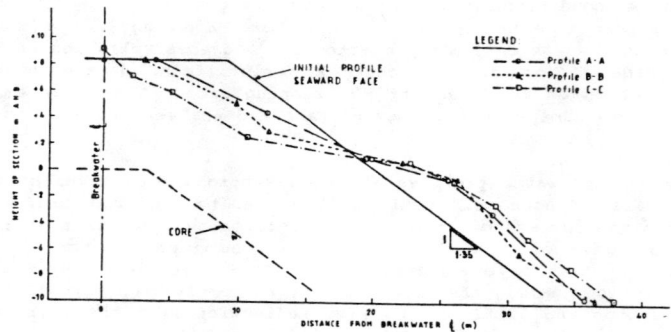

BREAKWATER CROSS SECTIONS — 2D FLUME TESTS
INITIAL AND FINAL PROFILES — MONOCHROMATIC WAVES
Crest Width = 18.0 m at R.L 8.39 m A.H.D.
4T Armour, Period 7 sec.

FIGURE 10

Previous tests carried out at the Water Research Laboratory on mass armour breakwater sections using both 2T and 4T armour units also have shown that the final equilibrium profiles are relatively independent of the mass of the armour units.

From the previous tests it was clear that there was significant overtopping of the structure although this did not result in any significant damage on the leeward face. To minimise the risk of damage to the leeward face by overtopping of the large waves in the group, the crest of the breakwater was raised by 2m to RL 8.39m AHD. To reduce the proportion of armour material in the breakwater, the core area was also extended.

Testing showed that the higher crest level reduced overtopping and splashing of water to the leeward face.

In order to investigate the feasibility of using 4T rock, tests were carried out on the same breakwater geometry. The linear scale for this test was 51.6. The final profile after 7 hours of testing and the stability of the structure and the equilibrium were similar to that for the 5T armour. The section as tested is shown in Figure 10.

To study the stability of the structure under random wave conditions, the same breakwater section was tested with random waves.

The tests were carried out for a duration of 7 hours using Pearson-Muskowitch spectra having a peak spectral period of 8 sec.

The breakwater was tested at three water levels, namely MSL, RL 3.75m AHD and RL 4.5m AHD which were the same as for the monochromatic wave tests. The characteristic significant wave heights of the spectra at the MSL and RL 3.75m AHD were 6m and 6.3m respectively. At the highest water level of RL 4.5m two spectra corresponding to H_s = 6 and 7.1 were tested.

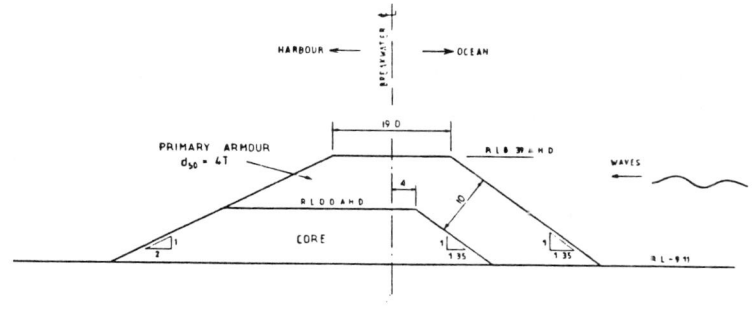

SECTIONAL GEOMETRY OF BREAKWATER
FINAL DESIGN   FIGURE 11.

Due to the depth limited conditions the larger waves in the group were breaking before reaching the structure.

The sectional geometry is nearly the same as that used in previous tests using 4T armour and 19m crest at RL 8.39m AHD. The layout of the breakwater is similar to the initial design except for realignment and extension of the breakwater head to reduce wave penetration into the harbour as shown in Figure 13.

The final design was tested in the 3D wave basin for waves approaching from the easterly direction. For stability tests on the breakwater a linear scale of 51.6 was used. At this scale the model armour material represents a d_{50} = 4T rock in the prototype. The topography of the ocean bed was not remodelled and was the same as that for the 60 scale model. The length of the breakwater was also based on a 60 scale. The sectional geometry of the breakwater, model water level, wave height and wave period were all based on 51.6 scale.

The breakwater (Figure 11) was tested at three water levels, MSL, RL 3.75m AHD and RL 4.5m AHD for waves up to 9m. For these tests the wave period used was 7 sec.

The final equilibrium profiles and the breakwater outline after 8 hours of testing clearly show that the extent of damage is limited to the seaward face. The variation of damage along the length of the breakwater was similar to that of the initial design which is described previously. Even at the section where worst damage took place, there was nearly 14m of crest which was unaffected by wave action. There was no significant overtopping and the leeward face was completely unaffected. The core was not exposed at any location along the breakwater. The tests clearly showed the structure to be stable even under the worst storm attack which it is likely to experience. (Figures 12 and 13).

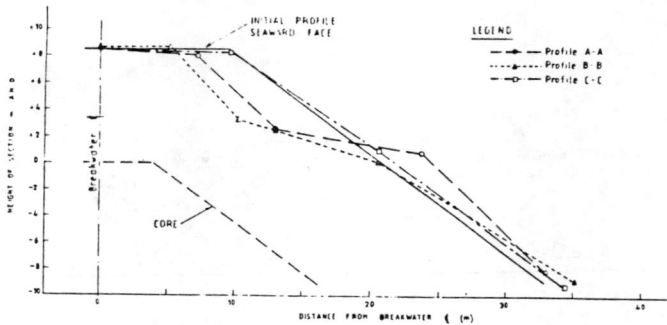

3D WAVE BASIN TESTS FINAL DESIGN
INITIAL AND FINAL PROFILES
Crest Width = 19.0 m at RL 8.39 m AHD
4T Armour, Period 7 sec
Wave Direction: East

FIGURE 12.

These model tests clearly showed that the mass armour breakwater was highly resilient and extremely stable when compared with a conventional breakwater using equivalent size primary armour. This increase in stability is attributed mainly to the high permeability of the structure which reduced drag and seepage forces as well as markedly reducing wave reflection. Wave heights inside the harbour for varying water levels are shown in Figure 14.

5.00 CONSTRUCTION

In 1983 investigations of the quarry at Mount Griffiths indicated the presence of massive rock in andesitic dykes. This led to a design of the breakwater by BBW, using a rubble mound mass armour rock structure which maximised the use of local material and which included large quantities of sound but heavily fractured rock mined in the process of extracting the required 4 tonne rock armour. The reject material from the quarry was to be used in the construction of a haul road 2.7km in length and a causeway 0.8km long.

Within the first eight weeks of the specified intensive development and operation of the quarry a much higher than predicted yield of fine material indicated that a considerable amount of 4 tonne nominal armour may need to be imported to construct the breakwater to the original design. From the first series of model test, it was believed there existed considerable scope to reduce the nominal size of the core material and possibly increase the volume of the core and still maintain a dynamically stable structure at the design wave and surge levels.

6.00 MODEL SERIES 2 DURING CONSTRUCTION

The purpose of this new series of tests was to examine options for including as much of the finer quarry run material as possible hence minimising, or even negating, the importation of extra material for the armouring of the breakwater.

ARMOURED BREAKWATER

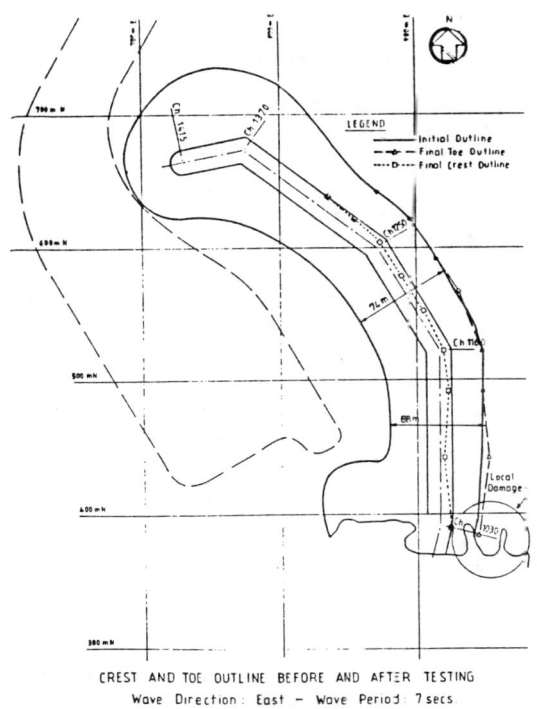

CREST AND TOE OUTLINE BEFORE AND AFTER TESTING
Wave Direction: East — Wave Period: 7 secs
FINAL DESIGN

FIGURE 13.

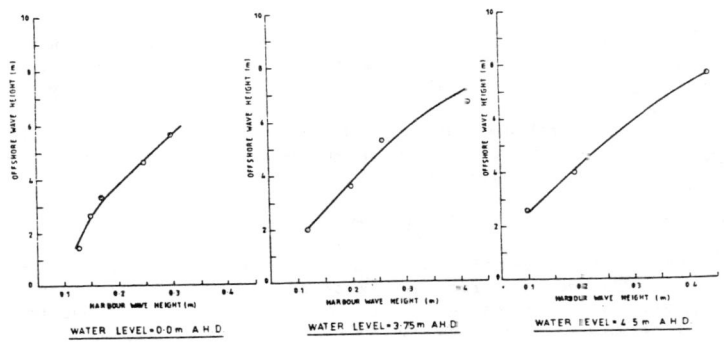

WAVE HEIGHT INSIDE THE HARBOUR FOR 7 SEC. PERIOD WAVES FROM EAST

FIGURE 14.

The report describes the two dimensional model flume testing of the following breakwater sections:
a) Breakwater trunk made entirely from the readily available quarry run material.
b) Core as per original design except that the nominal size is to be reduced from 2T to 1T.
c) Quarry Run Core. Core made completely out of quarry run material.
d) Composite Core. A core made up by substituting part of the nominal 1T core structure with quarry run material.
e) Composite Core Structure. The testing of the Composite Core covered by the nominal 4 tonne armour as in the original design.

Each of the breakwater sections was constructed in the test flume and the flume flooded to the appropriate depth. A one metre wave was applied and the water level varied over the normal tidal range (-2m to +2m) until all visible settlement of the model had occurred. The design storms were then applied to the structure. During this time a video film was taken at startup and shutdown of each stage of the test to record the model's performance. At the end of each stage of the design storms the average profile of the structure was recorded.

For the initial model test the water level was stepped up and down in discrete quantities to simulate the given water level for that design period of the storm. This method was abandoned for subsequent tests sections in favour of a more realistic linearly varying rise and fall of water levels to simulate storm surge.

The wave spectra as designed was input via the random wave flume computer. The test was run and periodic sampling of the spectra was made. A three probe analysis was made which divided the waves into incident and reflected. From this analysis the significant wave height and period as designed could be cross checked against that as recorded and minor adjustments made to bring the designed and recorded spectra into line.

The structure was tested against a 1 in 100 year storm based on a wave spectra emanating from a 950mb tropical cyclone approaching from an easterly direction and making landfall just north of Mackay.

The storm selected was designed to create the highest levels of damage to the structure by maximising the water level persistence at the 100 year level of 4.5m AHD.

6.01 Test A - Quarry Run Breakwater

This structure uses all available material (quarry run) with no armour on the seaward face and represents the investigation of a design extreme. The lee face is to be stabilised against overtopping using the originally proposed 4T nominal armour and the head to consist entirely of 4T armour.

This test sequence was run for general interest. The section shown in Figure 15 was tested in a 2D flume with random wave spectra of the same properties as used in the final tests in Series 1.

Grading tests on the quarry output at that date indicated the following size distribution.

Nominal Size (mm)	Percentage Passing	Nominal Weight (kg)
1000	100	2600
450	77	240
300	60	70
150	34	9
75	24	1
25	9	0

A comparison of the gradings prototype versus model are shown in Figure 16.

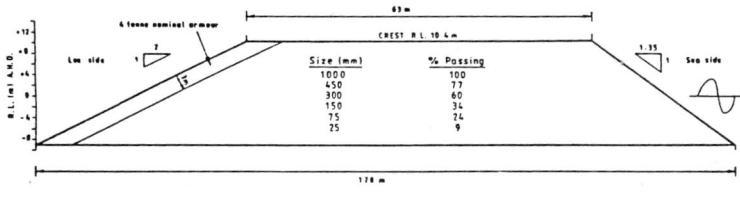

QUARRY RUN BREAKWATER DESIGN-TEST A   FIGURE 15.

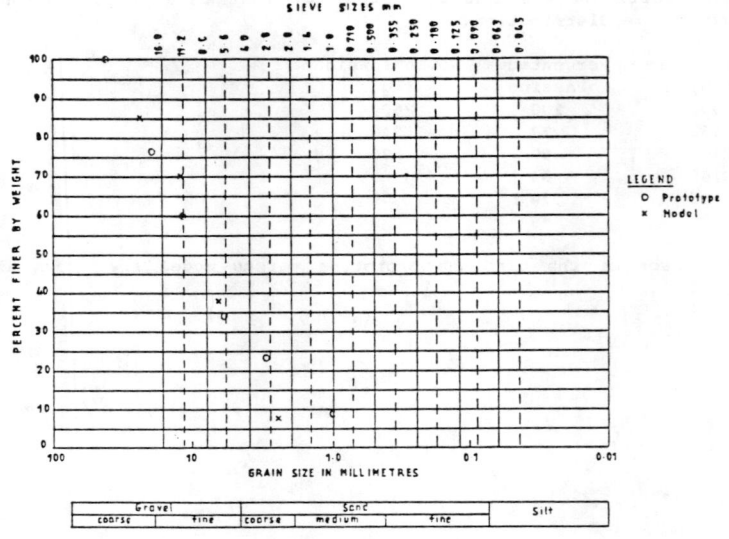

GRADING OF QUARRY RUN BREAKWATER (Model units)

Prototype (mm) = Model (mm) × 25

FIGURE 16

The model was subjected to the following sequence of storms:

a) Settling in - 1 metre waves over a number of tidal cycles.
b) First 100 year storm event.
c) Second 100 year storm event.
d) Summer storm - 2 metre waves over a number of extreme tidal cycles.
e) Estimated extreme 500 year storm event.

A time history of the profiles taken throughout the tests is shown in Figures 17 and 18. As expected massive reshaping of the structure occurred, however the long term profile demonstrates that the breakwater forms into a dynamically stable structure. The structure behaves like a beach where the profile acts to cause breaking of the waves before reaching the main body of the breakwater.

It is noted that the stepping of the storm water levels in the model has artificially built up berms. In the prototype, having a dynamic change in the water level throughout the storm, a more even profile would have resulted. It is concluded however that the berms have not affected the final profiles and that those given are considered representative.

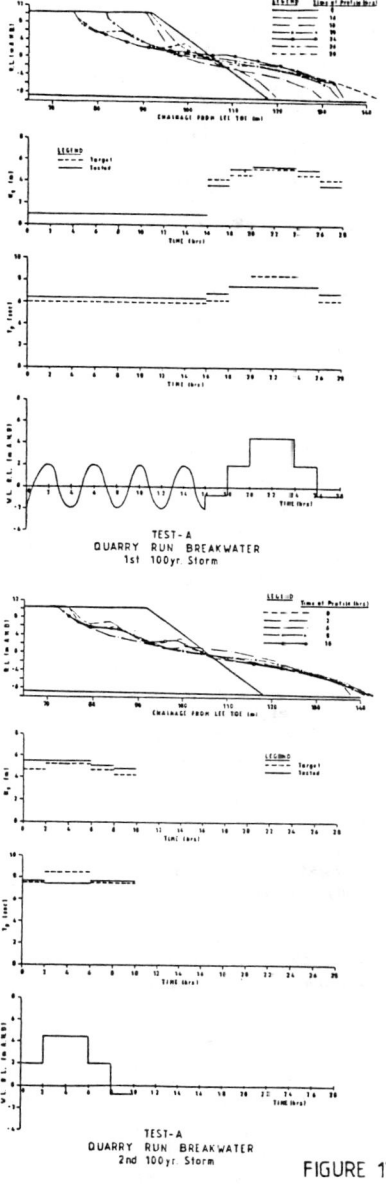

FIGURE 17

Tests B and C were carried out of core materials only with the core shape the same as the original design. Due to the shortage of armour and the possible time lapse involved in the obtaining or developing of other sources of armour rock, it was elected to test the stability of an unprotected breakwater core.

6.02 Test B: Core Reduced to Nominal 1 Tonne

The core was as per original design except using finer material, the nominal weight being reduced from 2 tonnes to 1 tonne. These tests were run to assess the risk of core construction being progressed without any armour.

The grading of the core is as follows:

Size Tonnes	Percentage Passing
2.0	100
1.0	50
0.25	0

The model was subjected to the following sequence of storms:
a) Settling in - 1 metre waves over a number of mean tidal cycles.
b) Winter storm - 2 metre waves over a number of extreme tidal cycles.
c) Summer storm - 3 metre waves over a number of extreme tidal cycles.
d) 100 year storm event.

A time history of the profiles taken throughout the tests is shown in Figure 18. It can be seen from the profile that during a normal winter storm there is minimal change to the core's profile. The larger waves progressively continue to flatten out the leading face until it becomes dynamically stable half way through the 100 year storm event.

6.03 Test C: Quarry Run Core

This design investigates the stability of a modified core design using only the quarry run material.

The grading size of this material is tabulated below. It was found that the quarrying methods in use would economically allow all larger rock sizes to be separately stockpiled and all fines rejected.

Size (mm)	Percentage Passing
450	100
150	0

ARMOURED BREAKWATER

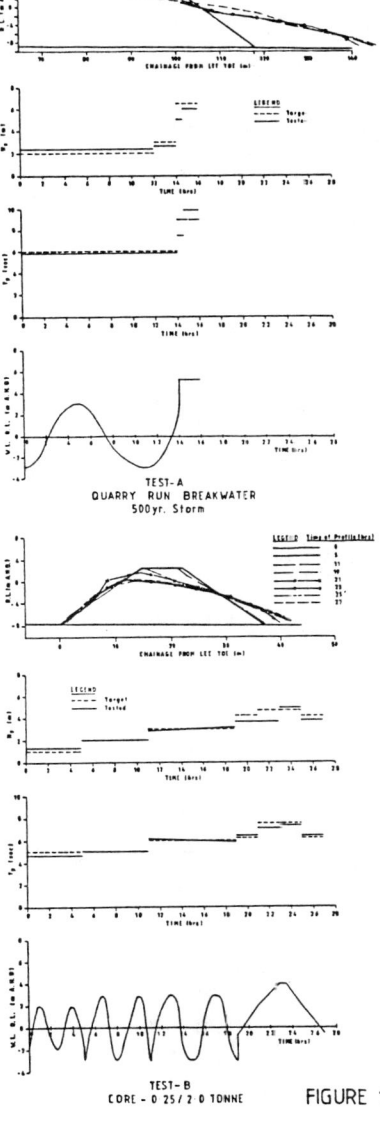

TEST-A
QUARRY RUN BREAKWATER
500 yr. Storm

TEST-B
CORE - 0.25/2.0 TONNE

FIGURE 18

178 BERM BREAKWATERS

The model was subjected to the following sequence of storms.

a) Settling in - 1 metre waves over a number of mean tidal cycles.
b) Winter storm - 2 metre waves over a number of extreme tidal cycles.
c) Summer storm - 3 metre waves over a number of extreme tidal cycles.

A time history of the profile is given in Figure 19.

As expected, massive reshaping to the structure occurred - even under settling conditions. As in the quarry run breakwater, the structure formed a leading protective beach.

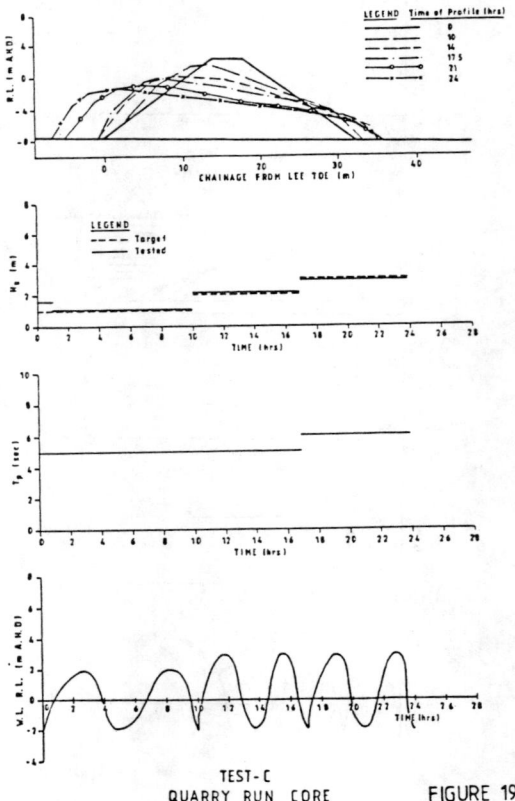

TEST-C
QUARRY RUN CORE   FIGURE 19.

ARMOURED BREAKWATER

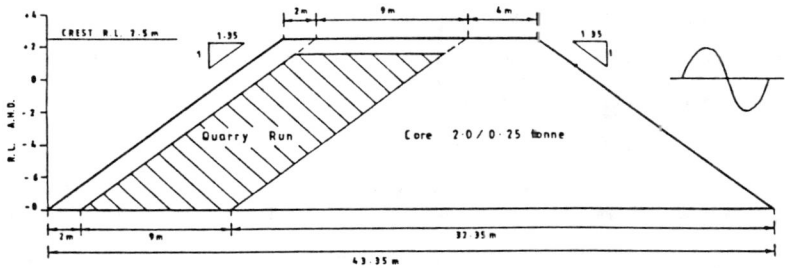

COMPOSITE CORE DESIGN TEST-D

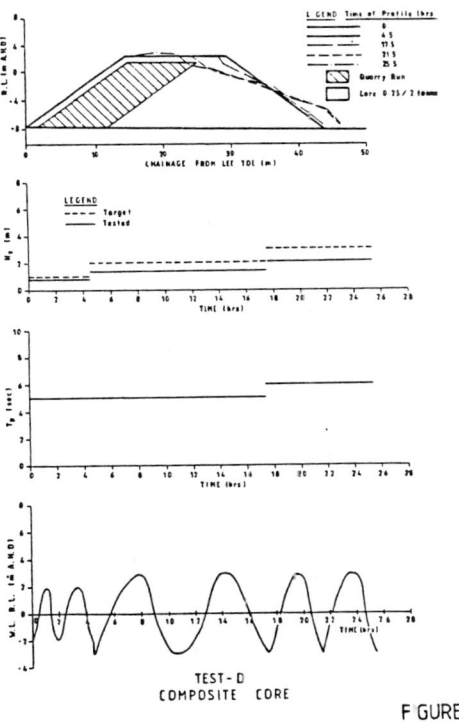

TEST-D
COMPOSITE CORE

FIGURE 20.

6.04 Test D: Composite Core

The core design and material is the same as that for Test B except that a section of the core material in the lee of the structure is replaced by quarry run material as shown in Figure 20.

The model was subjected to the following sequence of storms.

a) Settling in - 1 metre waves over a number of mean tidal cycles.
b) Winter storm - 2 metre waves over a number of extreme tidal cycles.
c) Summer storm - 3 metre waves over a number of extreme tidal cycles.

A time history of the profiles and storm conditions are given in Figure 20.

This structure performed in a very similar way to the structure containing no quarry run material.

6.05 Test E: Composite Core Breakwater

The composite core material was as described in Test D. The section tested is shown in Figure 21. A comparison of the model/prototype weight gradings of the armour are given in Figure 22.

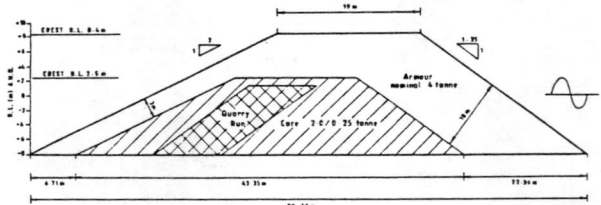

DESIGN OF COMPOSITE CORE BREAKWATER TEST-E   FIGURE 21

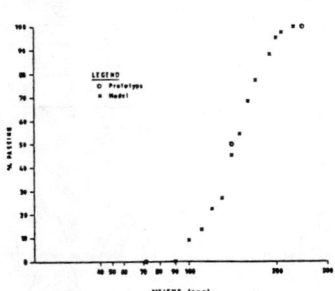

GRADING OF ARMOUR UNITS FOR COMPOSITE BREAKWATER
(Model units)
Prototype (g) = Model (g) = 28230

FIGURE 22

The model was subjected to the following sequence of extreme storm wave attack.

a) Settling in - 1 metre wave over a number of tidal cycles.
b) 100 year storm.
c) First 500 year storm waves with extreme water levels up to about a 2000 year return period.
d) Second 500 year storm waves with extreme water levels up to about a 2000 year return period.
e) Maximum waves. Largest monochromatic waves that will break continually on the structure for a wide range of water levels up to about a 2000 year period. To achieve maximum possible damage, wave heights and periods were adjusted so that maximum waves were made to plunge on the structure at all water levels.

A time history of the profiles and storm conditions are given in Figures 23 and 24.

The model tests clearly showed that this structure is extremely stable against a wide range of wave attack.

The model tests demonstrated that the potential existed for inclusion of the finer quarry material in the final design of the prototype and confirmed that large savings in construction costs are possible when this type of design and modelling approach is applied.

7.00 MODEL SERIES 3 - BREAKWATER HEAD DURING CONSTRUCTION

Following the results of test series 2 it was decided to further explore the stability of the breakwater head with a view to further reducing the armour quantity.

The final design of the breakwater shoreward of chainage 1330m was handed to the contractor on 2 June, 1986. The remaining outer length of the breakwater between chainage 1330m and the head at chainage 1413m was tested for the following modified designs:

i) Core Re-design
   a) Core cover to filter crest increased from 1.0m to 2.0m (i.e. RL 2.5m to RL 3.5m AHD).
   b) Core crest width increased from 17.0m to 24.0m.
   c) Seaward core slope steepened from 1:3 to 1:2.

ii) Armour Re-design
   a) Armour material grading on seaward slope reduced from 4-7t to 2-7t (4t nominal).
   b) Armour material grading on lee slope reduced from 4-7t to 0.25-2t (1t nominal), i.e. to the equivalent of the core material.
   c) Armour thickness over core crest reduced from 5.9m to 4.9m.
   d) Seaward armour slope steepened to 1:2.5 and 1:2 with the original 19.0m armour crest width maintained. Note that the original 1:3 sloped breakwater, although accepted from previous model work, comprises a thicker armour cover with a smaller core and therefore has also been tested for the modified design.

BERM BREAKWATERS

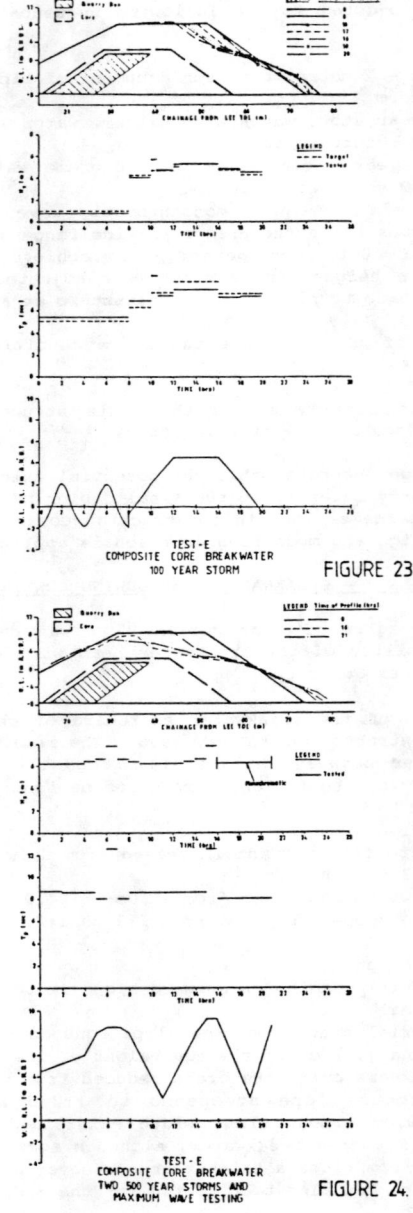

FIGURE 23. TEST-E COMPOSITE CORE BREAKWATER 100 YEAR STORM

FIGURE 24. TEST-E COMPOSITE CORE BREAKWATER TWO 500 YEAR STORMS AND MAXIMUM WAVE TESTING

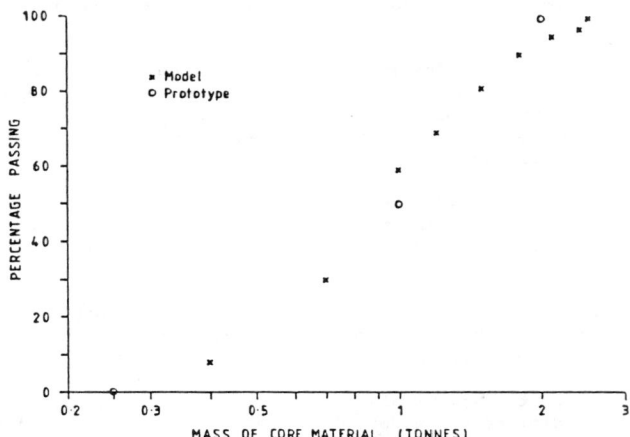

HAY POINT TUG HARBOUR
GRADING OF 0·25 - 2·0 TONNE CORE MATERIAL

FIGURE 25.

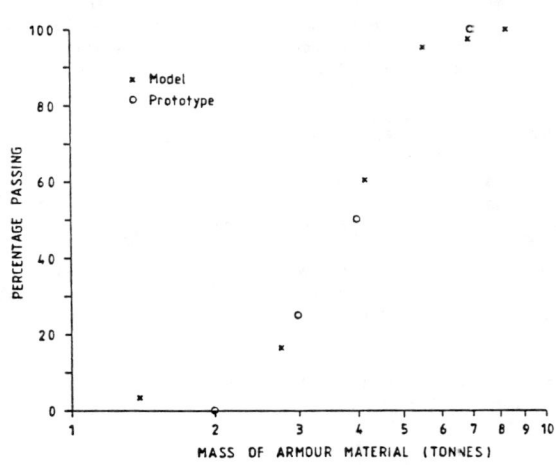

HAY POINT TUG HARBOUR
GRADING OF 2·0 - 7·0 TONNE ARMOUR MATERIAL

FIGURE 26.

The re-designed breakwater head defined above is presented in Figures 27 and 28.

In this testing programme of the re-design of the breakwater head, regular waves were generated such that wave breaking is initiated on the breakwater itself; the most severe wave attack scenario. The design water levels and associated incident wave heights used for testing were as follows:

Water levels: 0.0m to 4.4m AHD
Wave heights: 5.0m to 8.1m
Wave periods: 7.3s to 9.7s
Wave direction: N E

Breakwater stability tests were conducted in WRL's 30m long x 3m wide x 1.6m deep regular wave flume under fresh water conditions. Flume layout and survey codes of the model are shown on Figure 22.

Grading curves for the armour and core materials are given in Figures 25 and 26. The difference in rock densities together with the fresh-salt water buoyancy discrepancy is taken into account when establishing the breakwater material mass scale M_R.

Prior to testing the wave generator was calibrated. This involved establishing period and amplitude settings for the range of desired test water levels such that the most severe wave conditions were reproduced, i.e. wave breaking directly on the breakwater. The as constructed breakwater was then "bedded-in" under typical wave (H_s = 1 to 2m) and tidal (SWL = -2m to 2m AHD) conditions for an equivalent duration of 10 hours. Under such pre-test conditions, the breakwater is allowed to attain a denser, more compact structure which better represents the developed situation in the prototype.

Three Test Series comprising 15 tests were conducted altogether (i.e. 5 tests for each of the 3 breakwater configurations).

Test Series I   - Seaward head slope 1:2   - Test Nos. 1 to 5
Test Series II - Seaward head slope 1:2.5 - Test Nos. 6 to 10
Test Series III - Seaward head slope 1:3   - Test Nos. 11 to 15

For each Test Series the still water level was varied from approximately 0m to 4.5m and back down to 0m AHD for maximum breaking waves at the structure lasting 30 hours, i.e. 6 hours for each water level.

Table 1 sets out in summary form the equivalent prototype test condition for each test series.

Table 1 - Test Conditions

Test Series	Test No	SWL (m AHD)	H_B (m)	T (s)	Equivalent Prototype Test Conditions	
					Storm Duration (Hrs)	Total Test Time (Hrs)
I Seaward Slope 1:2	1	0.5	5.3	7.5	6.0	6.0
	2	1.5	6.6	9.7	6.0	12.0
	3	4.1	7.8	9.7	6.0	18.0
	4	1.6	6.6	9.5	6.0	24.0
	5	0.0	5.0	7.1	6.0	30.0
II Seaward Slope 1:2.5	6	0.0	5.4	7.3	6.0	6.0
	7	1.0	6.4	9.7	6.0	12.0
	8	4.4	7.4	9.7	6.0	18.0
	9	1.4	6.8	9.9	6.0	24.0
	10	0.2	5.9	7.4	6.0	30.0
III Seaward Slope 1:3	11	0.0	5.5	7.3	6.0	6.0
	12	1.5	6.5	9.7	6.0	12.0
	13	4.4	8.1	9.7	6.0	18.0
	14	1.7	6.6	9.7	6.0	24.0
	15	0.0	5.5	7.3	6.0	30.0

After the first 3 tests and following the final test within each test series, a survey of the breakwater was undertaken. Survey section locations are identified in Figure 27. In addition, after the third test within each test series (i.e. after testing at the maximum water level condition), the breakwater was photo-surveyed from above with the water level varied to identify the -2.2m, -0.2m, +1.5m, +4.5m and +6.5m AHD breakwater contours.

This structure is similar to the original design tested in test series 1 in both 2 dimensional flume and 3 dimensional basin model studies. The nominal weight (4T), crest level (8.4M AHD), crest width (19m), leeward slope (1:2), seaward slope (1:1.35) and thickness (10m) of armour are the same as the original design. Differences are present in the size of materials, crest level and leeward extent of the core.

A comparison of the profiles obtained after a 100 year event from this and series 1 tests showed that there are minimal differences between the profiles.

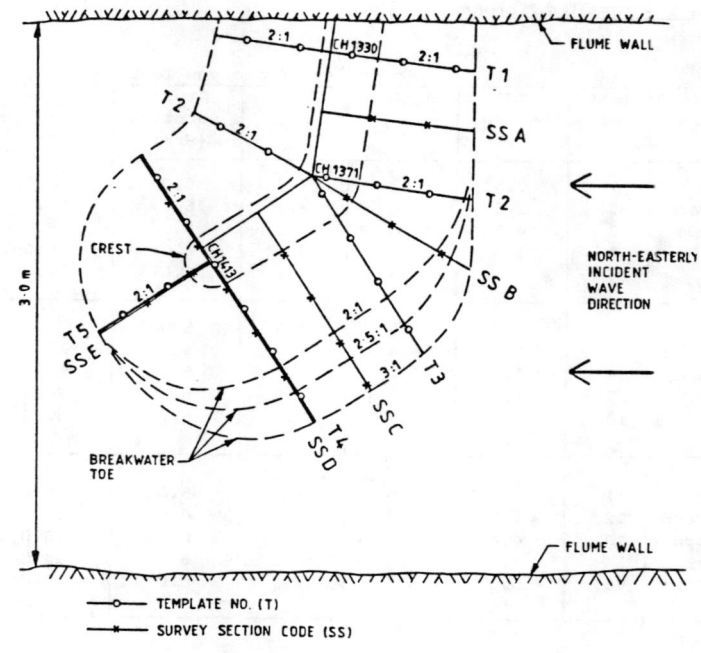

BREAKWATER MODEL LAYOUT FIGURE 27.

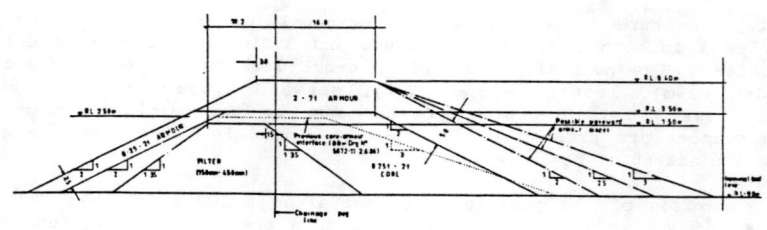

RE-DESIGNED BREAKWATER HEAD FIGURE 28.

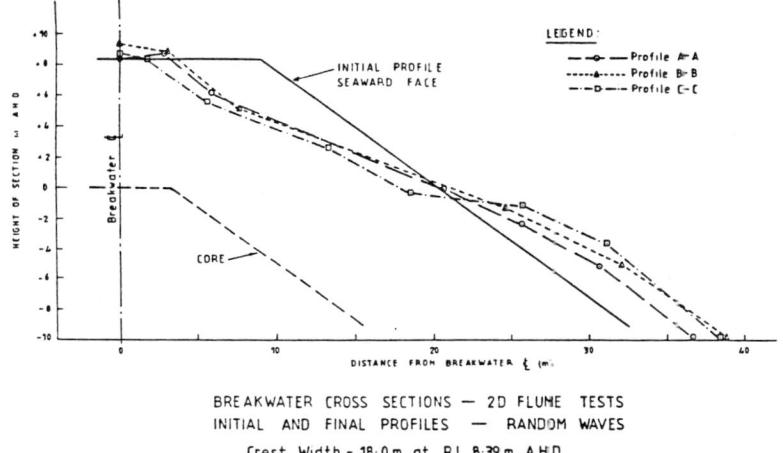

BREAKWATER CROSS SECTIONS — 2D FLUME TESTS
INITIAL AND FINAL PROFILES — RANDOM WAVES
Crest Width = 18·0 m at R.L 8·39 m A.H.D.
4T Armour, Period 7 sec.

FIGURE 29

A summarised video record of the 15 independent model tests has also been prepared.

Throughout all testing, no armour reshaping was observed to the lee slope of the breakwater. All reported reshaping or damage to the breakwater is confined seaward of the centre line. Representative surveyed results of the reshaped breakwater at section C for this test series are presented in Figures 30 to 32. All quantitative discussion pertaining to the reshaped structure must be deemed approximate and both chainage and elevation description taken as accurate to the nearest metre.

The results from Test Series III on the 1:3 slope head are in general agreement with previous extreme wave testing reported for waves from the East. The degree of reshaping and demonstrated stability of the overall structure were reproduced.

In all tests, irrespective of the degree of reshaping, or resultant armour over the core, a stable and resilient breakwater structure remained which effectively withstood the extreme conditions of breaking waves at high water levels imposed during testing.

BERM BREAKWATERS

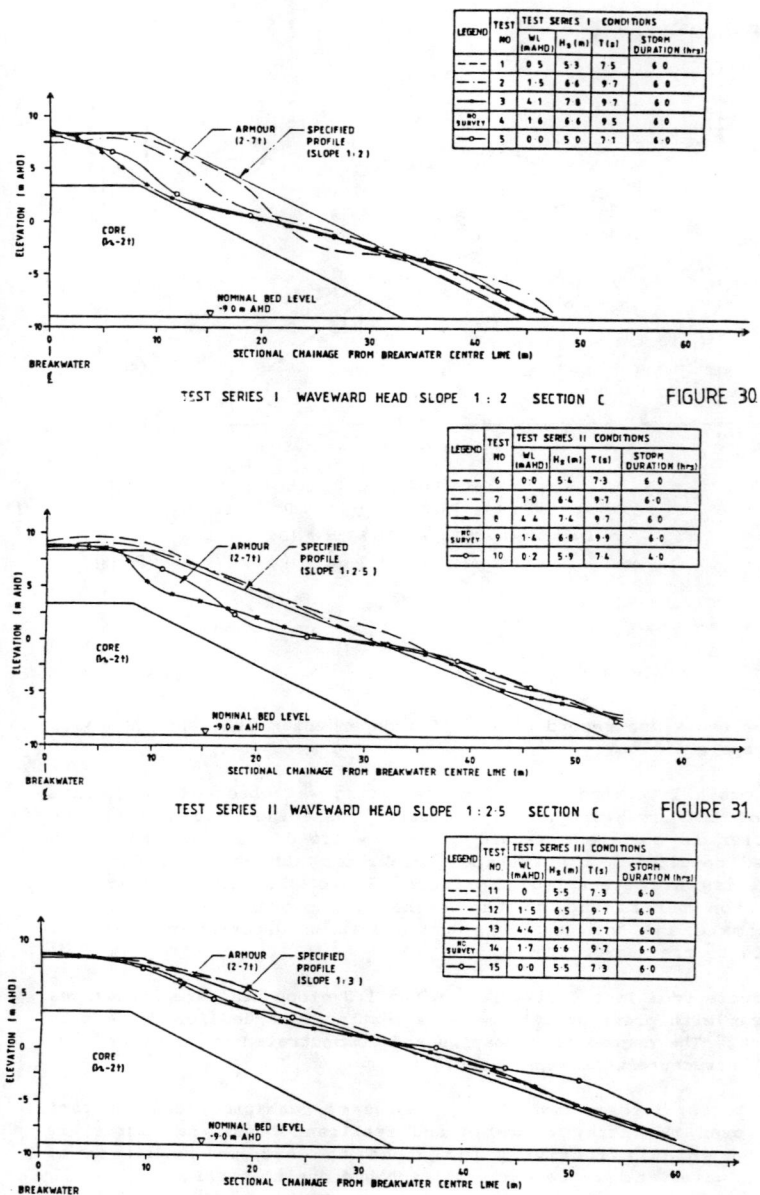

TEST SERIES I WAVEWARD HEAD SLOPE 1:2 SECTION C   FIGURE 30

TEST SERIES II WAVEWARD HEAD SLOPE 1:2·5 SECTION C   FIGURE 31.

TEST SERIES III WAVEWARD HEAD SLOPE 1:3 SECTION C   FIGURE 32.

ARMOURED BREAKWATER

Wave protection afforded to the harbour by the final reshaped structure is equal to that of the breakwater before reshaping occurred.

Because of the possibility that Mt Griffiths quarry may not produce sufficient armour to complete the job the schedule for tendering contained an item of 75,000t Breakwater Armour from a nominated source at Mt Scrubby which is about 36km from the site. Hence there was room in the contract to negotiate prices for armour rock from other sources. The contractor and the Principal conjointly developed an excellent quarry at Mt Chelona which is 32 km from the site and the breakwater was successfully completed using this source of armour rock.

The only other two existing quarries within economic range in the Mackay area in addition to being 42 km distant from the site also had a very low yield of 4t + armour rock.

The focus for redesign of the breakwater during construction concentrated on maximising the use of the local Mt Griffiths quarry output knowing that there would be a shortage of 4t nominal armour rock by the time this resource was exhausted.

Following the series 2 and 3 model tests drawings (Figures 33 and 34) for the final construction of the breakwater were issued in July 1986. Mt Griffiths quarry was closed on 17 September, 1986. The average rate of production was 3000 t.p.d. of all materials. Mt Chelona production averaged 700 t.p.d. of 4t to 7t and 1t to 2t armour stone of high quality. The lead from this quarry to the job is 36km.

The following tabulation of quantities used in this project are included to indicate its relative size.

Armour	4t to 7t	270,000 t
	1t to 2t	85,000 t
Filter		147,000 t
Core	1/4t to 2t	129,000 t
Causeway	Quarry Run Armoured with core material	850,000 t
	TOTAL	1,481,000 t

The breakwater, causeway and car park were completed on 18 March, 1987.

BERM BREAKWATERS

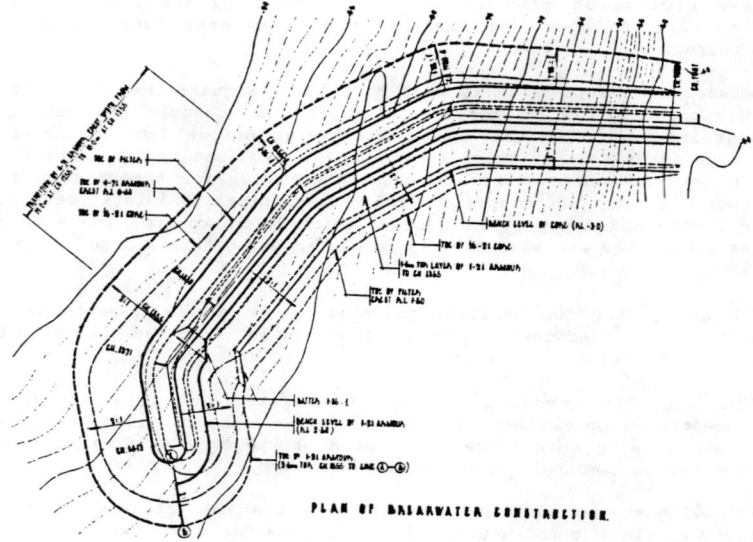

FIGURE 33.

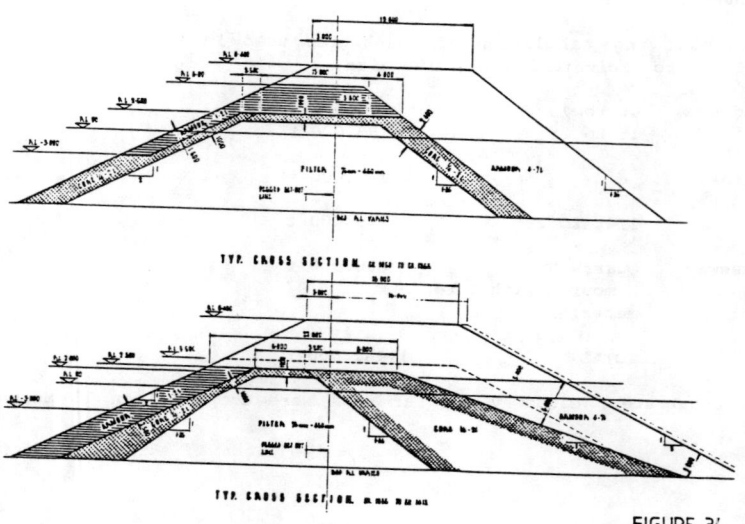

FIGURE 34.

8.0 DEVELOPMENT OF THE COMPUTER MODEL HARBREM

Previous Studies of Breakwater Stability

The vast majority of studies of breakwater behaviour have been based on experiment, due to the large number of variables affecting the problem and the lack of real knowledge surrounding the fluid behaviour. Generally, much of this work has been on an "as needs" basis for particular structures and as a result, the scientific yield has been small due to the absence of standardised procedures. A large bulk of "evidence" has therefore been collected over the years and correlated with the only available soundly based scientific study of the time - the Hudson Equation.

$$M = \frac{\rho_a}{K_D} \frac{H^3}{\Delta^3} \cot\alpha \tag{8.1}$$

where

M	is the mass of individual armour units:
ρ_a	the density of armour;
H	the height of the characteristic design wave;
Δ	the relative submerged weight of armour $(\rho_a - \rho_w)/\rho_w$
$\cot\alpha$	is the slope of the armour face

K_D is an empirical coefficient designed to account for the effects of all other (unknown) variables such as:

- armour type and shape
- number of layers
- armour placement method
- friction and interlocking
- water depth
- breakwater geometry
- size and porosity of underlayers
- wave spectra

Traditionally, results of breakwater tests have been expressed in terms of K_D, the so-called "damage coefficient", which has been reported to vary as a result from 1 to 150.

A major drawback of the Hudson Equation has been shown to be the omission of the effect of wave period in the basic equation. It is now recognised that wave period does influence breakwater behaviour because it controls, together with other factors, whether or not a wave of given height will plunge or simply surge against the structure. The difference in resulting wave forces is considerable, the breaking wave force being an order of magnitude greater than the non-breaking wave.

In search of an alternative approach which might be better suited to the mass armour design the work by Meer was examined and found to offer certain advantages.

Basis Of The Meer Empirical Model

Development of the Meer Model progressed from the reanalysis of other researcher's results such as Ahrens, Losada et al and Thompson and Shuttler. In particular, a cornerstone of the Meer model is the correlation of stability as a function of the Surf Similarity Parameter ξ_z after Battjes. This single parameter embodied the essential interaction between wave period and armour slope needed to correlate the previous test data and allowed Meer to design an extensive series of physical model tests to further support the concept. The other major contribution was in establishing a clear correlation of damage against duration of wave attack.

The following variables have been identified as significant to the breakwater design problem.

Nominal Armour Diameter

$$D_{n50} = \left(\frac{W_{50}}{\rho_a}\right)^{1/3} \tag{8.2}$$

where D_{n50} = nominal armour diameter
W_{50} = actual 50% value of armour mass distribution
ρ_a = mass density of armour

Relative Mass Density of Armour

$$\Delta = \rho_a/\rho_w - 1 \tag{8.3}$$

where ρ_w = mass density of water (salt or fresh)

Armour Slope

Implicitly this variable (like armour mass) is a large factor in design, with flat slopes ($\cot\alpha > 2$) generally exhibiting much increased stability over steeper slopes. However, total structure cost is also highly dependent on the choice of this parameter.

Wave Height

The incident wave height is the most commonly used indicator of wave energy, although it is used in various forms. The Hudson Equation uses a "characteristic height" H which normally includes the reflected wave superimposed on the incoming wave and thereby also contains information perhaps more correctly attributed to structure "permeability". In the present study, following Meer, random waves derived from a Pierson - Moscowitch (P-M) spectrum are used and the incident wave is characterised by the "significant wave height" H_s.

Wave Period

$$T_z = 0.71 \, T_p \tag{8.4}$$

where T_p = period of peak spectral energy

Hudson Stability Number

$$N_s = \frac{H_s}{\Delta D_{n50}} \quad (8.5)$$

where N_s = the "stability number" or "normalised wave height".

The Hudson Equation can then be rearranged such that

$$N_s = (K_D \cot\alpha)^{1/3} \quad (8.6)$$

where K_D = the "damage coefficient"

Surf Similarity Parameter

This single dimensionless parameter provides a valuable measure of the combined effects of wave height, wave period and armour slope acting externally on the structure;

$$\xi_z = \frac{\tan\alpha}{\sqrt{2\pi\ H_s/(gT_z^2)}} \quad (8.7)$$

This allows classification of the incident wave regime, based on experiment, such that for

ξ_z < 2.5 - 3.5; waves will tend to plunge on the structure
ξ_z > 2.5 - 3.5; waves will tend to surge against the structure
2.5 < ξ_z < 3.5; an intermediate condition exists.

The Meer model predictions are best summarised in terms of N_s versus ξ_z.

Unit Damage Level

The basic correlation in the Meer model is related to the eroded area of armour A_e, below the original starting profile, but this is extended to a non-dimensional form as the unit damage level S, viz

$$S = \frac{A_e}{D_{n50}^2} \quad (8.8)$$

Physically, S is then the number of cubical stones of size D_{n50} eroded over a unit width D_{n50}.

Structure Permeability

Definition of this parameter has been universally elusive. Meer was forced to carry it through as a dimensionless coefficient P, similar in effect to the Hudson K_D value. The present study aimed to establish a more rational basis for this parameter based on the actual armour layer thickness D_a.

Duration of Wave Attack

This simple concept and its obvious application to the problem of breakwater stability was fully exploited by Meer through a systematic approach to model testing.

The duration of attack is directly measured as the number of waves N, on the basis of the average zero crossing period T_z. Profiles of A_e were taken at regular intervals of N waves.

Other Factors

Meer also touched on other aspects such as spectral shape and wave groupiness, both thought to embody significant although secondary effects. Armour mass distribution (e.g. uniform stones versus "riprap") is another factor which affects void ratio and hence "permeability" which was partly investigated by Meer and will need more attention by future researchers.

Still water level (SWL) is another factor which obviously limits the height of the incident wave but is also known to influence wave runup. This was kept constant in the Meer investigation.

The range of permeability values P was derived by curve fitting such that:

$$0.1 < P < 0.6$$

was determined by Meer for the structures used.

Unfortunately Meer did not state the thickness of the uniform armour structure used in the type A structure tests. Overall, the tests were modelled on the "conventional" designs where the bulk of experience and test results are available.

Meer also concluded that spectral shape and groupiness of the wave train had no measurable effect on stability for the range of tests conducted. Likewise the differences between uniform stones and riprap were reported as immeasureable.

The influence of duration of attack to stability was found to be strongly correlated such that $S \alpha \sqrt{N}$.

Meer Stability Formulae

Two distinct types of structure behaviour were isolated and categorized according to ξ_z;

a) for plunging waves ($\xi_z < 2.5$ to 3.5)
$$N_s = 6.2 \ P^{0.18} \ (S/\sqrt{N})^{0.2} \ \xi_z^{-0.5} \tag{8.9}$$

b) for surging waves ($\xi_z > 2.5$ to 3.5)
$$N_s = 1.0 \ P^{-0.13} \ (S/\sqrt{N})^{0.2} \ \xi_z^{P}\sqrt{\cot\alpha} \tag{8.10}$$

Within the transitional zone ($2.5 < \xi_z < 3.5$) the results from both Equations must be compared to determine which solution prevails.

General conclusions embodied in the above equations are that minimum stability occurs around the transition from plunging to surging. For plunging conditions ξ_z describes the influence of both wave period and slope angle whereas for surging conditions different results are found for each slope angle.

In addition, for impermeable cores, wave period effects for surging waves are small but are more evident in the case of permeable cores, comparable to the effect of plunging waves.

Difficulties in Applying the Meer Model

The first major difficulty with this model is its complexity, at least when compared with the Hudson Equation. This is an unavoidable by-product of increased sophistication but the equations are nevertheless somewhat daunting at first encounter, especially since ξ_z is itself a function of other variables. Also, both equations must be solved in the transitional zone which leads to further possibility of error.

Incorporation of the various equations into a computer program is almost an essential adjunct to its general application.

Another problem relates to the choice of S as the indicator of damage which, even as Meer acknowledges the extent of damage depends on the slope angle. More stones have to be displaced for gentler slopes before the "failure" criterion is reached. On this basis, and for the tests conducted, S values corresponding to "failure" ranged from 8 to 17. The present study addresses this problem by extending the unit damage level to a layer damage level which accounts for the slope angle.

The structure types tested are also difficult to interpolate and/or extrapolate in terms of deciding on a representative P value for configurations where more than two layers may be desired. The type A (uniform, no core, no filter) structure is not usefully applicable because the thickness of the armour layers is not given. The present study worked towards establishing a more rational estimator for P by exploring the relative influences of different numbers of armour layers.

Finally, the Meer results address the question of so-called "static" stability of the structure where it is assumed little change in profile shape occurs. Of more interest in the present study was the added effect of quite large changes in profile shape, or "dynamic" stability, where reshaping of the structure is not only expected but in many cases desirable. All of the above aspects were considered when designing the series of model tests described in the following sections.

PHYSICAL MODEL TESTING

The primary aim of the tests was to investigate the changes in structure behaviour as a function of the thickness of the armour layer. Incident wave conditions were allowed to vary only over a limited range such that wave steepness was near constant and plunging waves predominated (near the most critical stability condition). Other parameters such as water depth and armour slope were kept constant throughout the tests to simplify understanding and interpretation of the results.

TABLE 2 - SUMMARY TEST PARAMETERS

	D	W_{50}	D_{n50}	Armour Thickness Test Series			Incident Wave Conditions					
							Wave A		Wave B		Wave C	
				1	2	3	H_s	T_p	H_s	T_p	H_s	T_p
Prototype Units	15.0	4500	1.2	11.7	7.8	3.9	4.0	6.0	5.0	7.0	6.0	8.0
Model Units	.50	.160	.0385	.39	.26	.13	.13	1.10	.17	1.28	.20	1.46

(All units kg, m, s)

The selected experimental breakwater structure is shown in Figure 35. Thickness of the armour layer was varied as three discrete values, referred to as Series 1, 2 and 3. Still water level was maintained constant at 0.5m at the toe of the structure. A series of three wave conditions (A, B and C) were applied to each armour layer thickness configuration for a total of nine separate tests.

Test Procedures

A nominal length scale (L_R) of 30 was used in formulating the model parameters, based on Froude scaling criteria. Table 2 summarises the model/prototype conditions tested.

Testing was performed in the programmable random wave flume of the N.S.W. Department of Public Works Hydraulics Laboratory at Manly Vale in Sydney. The facility is outlined in Figure 36 as being 1m in width, 1.5m deep and with overall length 30m. The flume paddle motions were computer controlled and water levels were logged from three capacitance wave probes to enable discrimination between incident and reflected wave energy.

Series 1 tests were performed first, subjected to each of the three sets of wave conditions but rebuilt and raked over prior to each wave change. Some armour stone was then removed to conduct Series 2 and Series 3 tests.

Prior to the commencement of each test a reference profile was recorded, against which to measure subsequent armour damage. Pierson-Moskowitch (P-M) spectral forms were used to characterise the random sea conditions.

The duration of each test was 5,000 waves, based on the average zero crossing period T_z (in line with Meers). After each 1,000 waves the tests were paused and a single centreline armour profile was taken. Maximum levels of wave runup (R_u) and rundown (R_d) were taken visually, measured against the glass sidewall of the flume.

Discussion of Results

A typical set of progressive armour erosion profiles is given in Figure 37 showing gradual development of the characteristic "S-shaped" curves, with the formation of a flattened berm area extending out to around the limit of wave rundown. Use of identical wave paddle control sequences ensured good repeatability of H_s and T_p across all nine tests. Runup and rundown, somewhat surprisingly, was found to vary only slightly with armour thickness and appeared unaffected by the progressive profile changes throughout an individual test. The reflection coefficient behaved in a likewise manner, being predominantly a function of the incident wave alone.

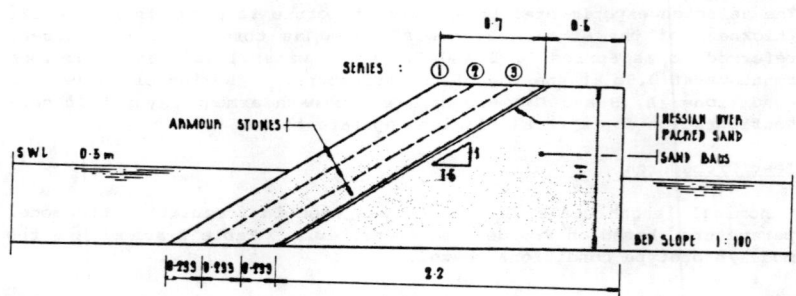

EXPERIMENTAL BREAKWATER STRUCTURE

FIGURE 35.

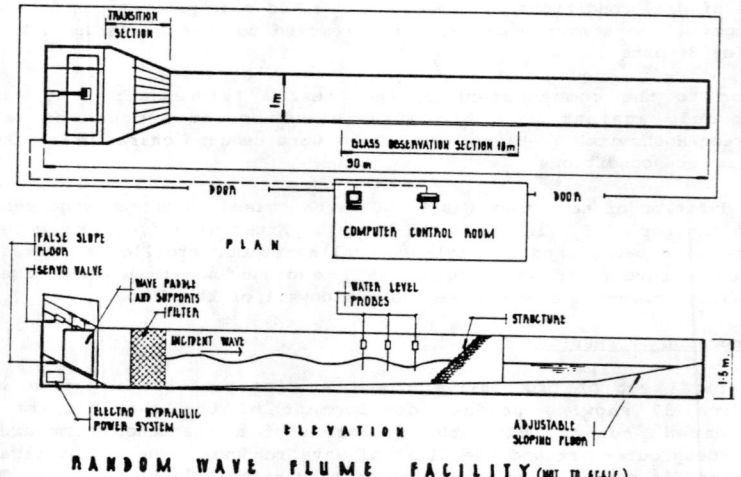

RANDOM WAVE FLUME FACILITY (NOT TO SCALE.)

FIGURE 36.

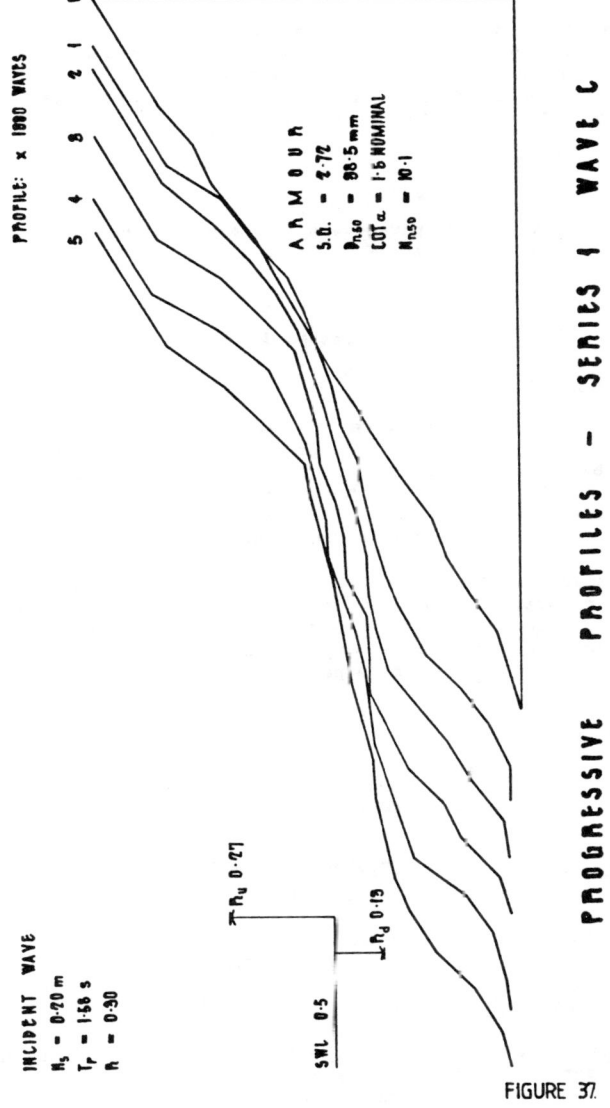

FIGURE 37.

Damage Criteria Definitions

Figure 38 details the major variables used in describing the structure behaviour. Following Meer, the eroded area A_e below the original profile $\cot\alpha$ is the primary damage indicator, found to be well represented by a sine curve over an eroded base length L_e and with maximum eroded depth D_e; viz

$$A_e = \frac{2}{\pi} D_e L_e \qquad (8.11)$$

New dimensionless parameters are also defined, being the relative armour layer thickness;

$$N_{n50} = D_a/D_{n50} \qquad (8.12)$$

and the maximum relative eroded depth;

$$\delta\Delta_{n50} = D_e/D_{n50} \qquad (8.13)$$

The eroded profiles were analysed to determine A_e and D_e in each case. From these, the damage level S was calculated, together with L_e based on Equation 8.11. This method of calculating L_e from D_e rather than as a direct measurement was found to be a practical way of disregarding relatively minute surface armour layer shifts at the coarse level of profile definition.

Although only a small number of tests were performed, there were clear responses for the increase in damage as a function of wave type, duration N and relative armour thickness N_{n50}. In general, lower levels of damage were sustained by the thicker armour layer (Series 1), as would be expected.

Figure 39 shows the increase in layer damage δ_{n50} versus duration N for each test.

Values of Meer's "permeability" coefficient P were then derived for each test result, based on the plunging regime formula (Equation 8.9), i.e.

$$P_{best} = \left[\frac{6.2}{N_s} \frac{(S/\sqrt{N})^{0.2}}{\sqrt{\xi_z}}\right]^{-1/0.18} \qquad (8.14)$$

and based on these values, Figure 40 shows P_{best} as a function of N_{n50} and the adopted line of best fit as

$$P = 0.017 N_{n50} + 0.044 \qquad (8.15)$$

derived using a least squares approach, but excluding outlier values for test 2A, assumed to be the result of some localised slippage of armour during early stages of the test. The only independent Meer reference point is at (2.2, 0.1) which lies above the line in this case, the other Meer values (P = 0.5, 0.6) being less easily located on this plot due to lack of interpretation and information respectively.

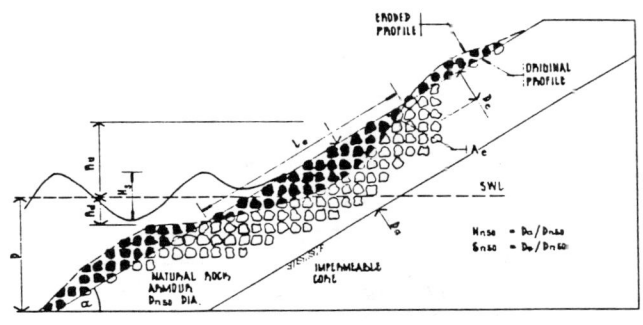

DAMAGE CRITERIA DEFINITIONS

FIGURE 38.

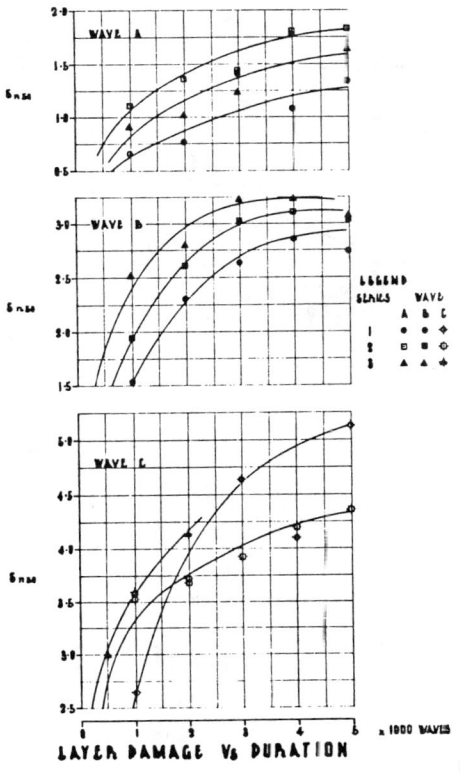

LAYER DAMAGE Vs DURATION

FIGURE 39

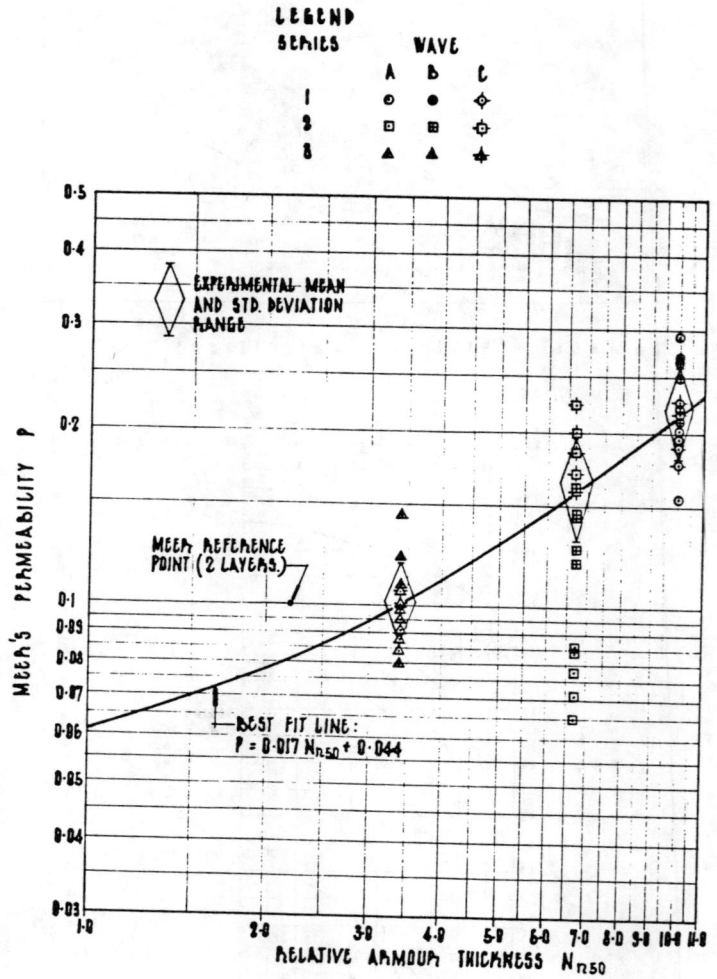

ADOPTED $P - N_{n50}$ RELATIONSHIP

FIGURE 40

Using these values, back substitution into the plunging regime formula yields values of the independent variables shown plotted on Figure 41 with dependent variable $S/\sqrt{N}$ and overlaid by the Meer best fit relationship. On this basis the plunging regime fit remains very reasonable.

Estimates of Length of Eroded Slope

The suitability of a "layer damage" model extension to the Meer "unit damage" model relies on an independent estimate for the eroded slope length L_e. A seemingly rational basis for establishing the scale of this parameter is the swash zone length, i.e.

$$L_e = f (R_u + R_d)$$
$$= f (H_s, \xi_z, P) \text{ intuitively} \qquad (8.16)$$

Assuming the extent of A_e is relatively small and the effect of changes in $\cot\alpha$ is also small, then $(R_u + R_d)$ might be expected to be near constant for a given geometry. In fact, not only was $(R_u + R_d)$ insensitive to A_e throughout any given test, it was also largely insensitive to P (or N_{n50}). A fit to the relationship between L_e and $(R_u + R_d)$ was based on equating the parallel-to-slope swash zone to the eroded area base, viz

$$L_e = k (R_u + R_d) \sqrt{1 + \cot^2\alpha} \qquad (8.17)$$

where k = 1.2. This model then estimates L_e as 20% larger than the swash zone length.

THE PREDICTIVE MODEL

The model (HARBREM) has been formulated to simplify the design process by presenting the engineer with a broad view of the problem and at the same time using parameters which are easily understood and assimilated. By showing a range of predicted structure behaviours the designer can immediately gauge the areas of sensitivity and avoid them, opting for a more "plastic" response region.

The present model reflects the traditional "design" approach, i.e. the design storm is chosen "a priori", being f(environment). Then a particular geometry ($\cot\alpha$) is chosen and the design problem reduces to armour selection and the questions of "How heavy?" and "How many layers?" are answered on the basis of the predicted degree of erosion. Also, because rubble breakwater design depends heavily on economic considerations, the model aids the designer in choosing the overall least capital cost solution.

The extension of the Meer formulae to address armour layers, rather than unit damage alone, relies on a good independent estimate of the eroded area base length L_e. With a correlation established relating L_e to (R_u+R_d), it leaves then an estimate for the latter to be obtained. Perhaps one of the most comprehensive summaries of runup and rundown has been presented by Losada et al where six sets of R_u and four sets of R_d are correlated against the surf similarly parameter ξ_z. Each experimental data set was fitted to an equation of the form.

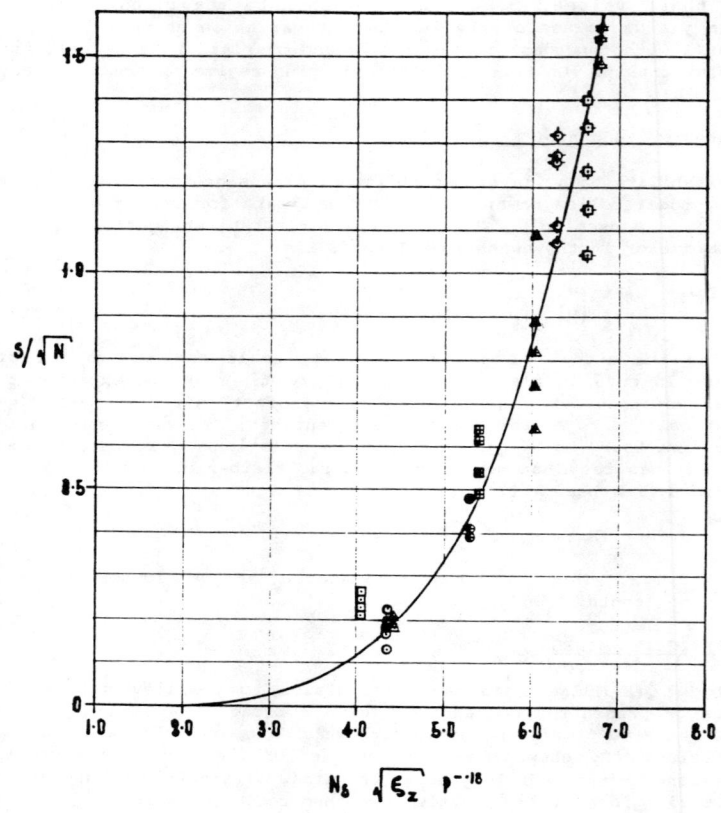

FIGURE 41. CURVE FIT TO PLUNGING REGIME STABILITY FORMULA ($\xi_z < 2.5-3.5$)

$$R_u/H = A(1 - e^{B\xi_z}) \qquad (8.18)$$

where H is the characteristic wave height and A and B are curve parameters. Figure 42 shows the experimental results from the present study and the Losada best fit curve for a quarrystone slope tested by Kamel and Dai. The resulting approximate parameters are:

	A	B
R_u	1.4	-0.4
R_d	-0.7	-0.4

The following development then steps through all the necessary calculations to arrive at the minimum stable armour size for a given "design" condition.

Given Variables

a) Design storm = f(environment):
  H_s;   significant wave height   (m)
  T_z;   zero crossing period   (s)
  t_d;   storm duration   (hr)

b) Geometry details:
  $\cot\alpha$;   slope of breakwater face
  N_{n50};   relative armour layer thickness
  δ_{n50};   maximum relative eroded layer depth

c) Physical details:
  ρ_a;   armour density (tonnes/m^3)
  ρ_w;   water density (tonnes/m^3)

Calculated Variables

$\Delta = f(\rho_a, \rho_w)$
$\xi = f(H_s, T_z, \cot\alpha)$
$P = f(N_{n50})$
$L_e = f(R_u, R_d, \cot\alpha, k)$
$\quad = f(H_{max}, \xi_z, A, B)$

where $H_{max} = H_{ratio} H_s$
with normally H_{ratio} = 1.86 for 0.1% exceedance of Rayleigh distribution.

The minimum stable armour nominal diameter D_{n50} is obtained from:

$A_e = f(D_e, L_e)$
$S = f(A_e, D_{n50})$
$\delta_{n50} = f(D_e, D_{n50})$

with simultaneous solution of the above;

$$D_{n50} = \frac{2}{\pi} \frac{\delta_{n50} L_e}{S} \qquad (8.19)$$

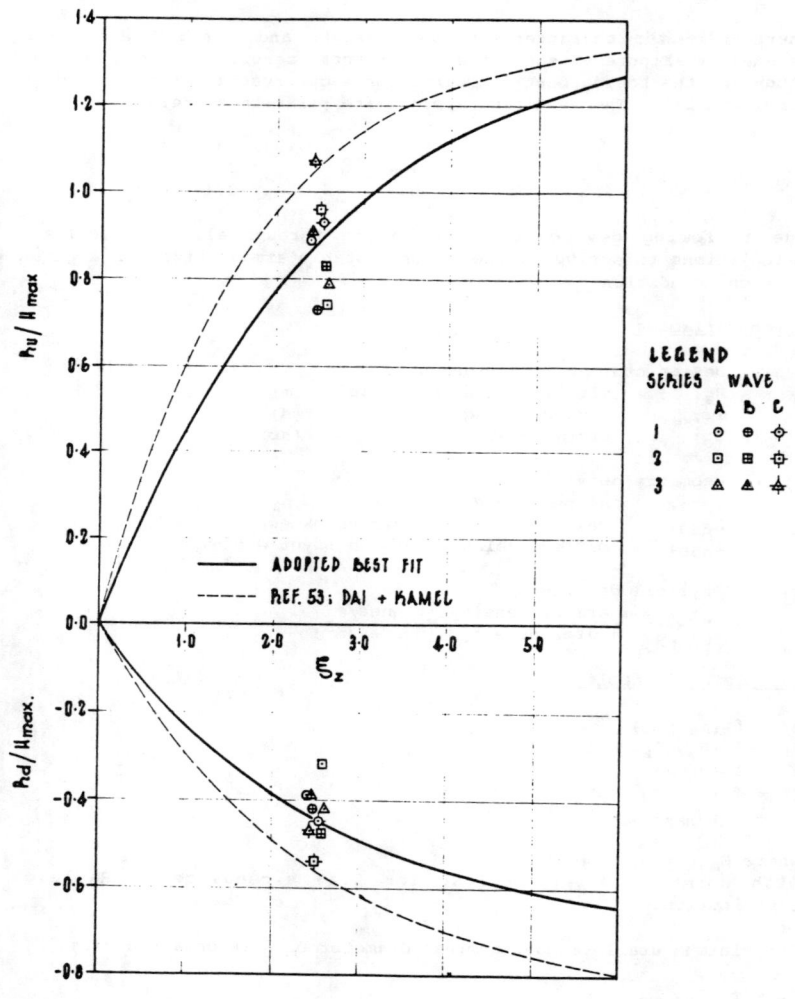

COMPARISON OF RELATIVE WAVE RUN-UP AND RUN-DOWN VALUES

FIGURE 42.

where S is derived from the Meer formulation, but firstly by considering the stability equations in their (generic) damage and duration independent forms, i.e.

$$D_{n50} \left(\frac{S}{\sqrt{N}} \right)^{0.2} = f(H_s, \xi_z, \Delta, P, \cot\alpha) = D'_{n50}$$

so that

$$S = \sqrt{N} \left(\frac{(D'_{n50})}{D_{n50}} \right)^5 \tag{8.20}$$

Substituting into Eq 8.19 gives,

$$D_{n50} = \left[\frac{\pi}{2} \frac{\sqrt{N}}{\delta_{n50}} \frac{D'_{n50}{}^5}{L_e} \right]^{1/4} \tag{8.21}$$

where D'_{n50} is given by either, for plunging wave conditions:

$$(D'_{n50})_p = \frac{H_s \sqrt{\xi_z}}{6.2 \; P^{0.18} \; \Delta} \qquad \text{for } \xi_z < 2.5 - 3.5 \tag{8.22}$$

or, for surging wave conditions:

$$(D'_{n50})_s = \frac{H_s}{1.0 \; P^{-0.13} \; \sqrt{\cot\alpha} \; \xi_z{}^P \; \Delta} \qquad \text{for } \xi_z > 2.5 - 3.5 \tag{8.23}$$

The model calculates D_{n50} on the basis of both the above equations and chooses the minimum of the two solutions as the correct value. Other derived parameters may then be calculated, such as

$$W_{50} = f(\rho_a, D_{n50})$$
$$N_s = f(H_s, \Delta, D_{n50})$$
$$K_D = f(N_s, \cot\alpha)$$

An estimate of design capital cost is then achieved by considering the armour mass per unit surface area normal to the armour slope.

$$M_A = N_{n50} \; D_{n50} \; \rho_a \; (1 - \phi) \tag{8.24}$$

where ϕ is the average armour porosity (assumed constant 0.4 by default) and the cost per unit area, similarly

$$C_A = M_A \; C_{n50} \tag{8.25}$$

where C_{n50} is the estimated $ cost per tonne of armour mass W_{50} quarried, transported and placed on the structure. Given a range of possible N_{n50} and $\cot\alpha$ values the model selects the lowest cost solution from the alternatives.

Finally, the estimated total breakwater armour cost for the lowest cost (N_{n50}, W_{50}) solution is

$$B_C = B_L \ B_H \ \sqrt{1 + \cot^2 \alpha} \ C_A \qquad (8.26)$$

where B_L and B_H are length and crest height of structure respectively.

A calibration check of the model was performed by back-substituting into the equations with the aim of deriving calculated values of S, δ_{n50} and L_e as a function of H_s, T_z and t_d; given the geometric properties of the structure. In this way the approximations implicit in the definition of the P-N_{n50} relationship and the $(R_u+R_d) - L_e$ relationship would become evident.

Also to provide an independent check of the model performance a similarly designed and conducted set of test results was needed. Unfortunately this was difficult to obtain because of the wide variety of tests undertaken during the earlier Half Tide investigations. These were targeted at specific design conditions and in particular had quite high changes in SWL and/or considerable overtopping of the model structures occurred. Only one test (Test E, Figures 21 and 23) was a close approximation to the structure of Figure 35.

The comparisons between unit damage measurements and predictions is given in Figure 43. It should be noted the model S value prediction is not a function of the L_e prediction. These show a reasonably good agreement, generally giving a prediction within 30% of the measured value for both calibration and verification data. The exceptions to this rule are at the low damage end where the model tends to underpredict damage. Generally however the result is considered a good one, taking into account the limited scope of the testing.

Finally, Figure 44 shows the results of the δ_{n50} layer damage predictions versus measured, with the values now being a function of L_e. In spite of the errors in L_e, again the prediction is of the order of 30% of measured values but with a more pronounced tendency to under-predict at low damage levels and over-predict at high damage levels. The over-prediction of eroded layer depth at high levels can be directly related to the inability of the present model to allow an increase in L_e with time.

EXAMPLE MODEL USAGE

The model is easy to use, fast and only requires a printer for output. Input can be either in interactive mode or via a prepared input file. Any number of separate design cases can be examined in a single model run.

Input

The example treated here is broadly based on the design of the Half Tide harbour breakwater.

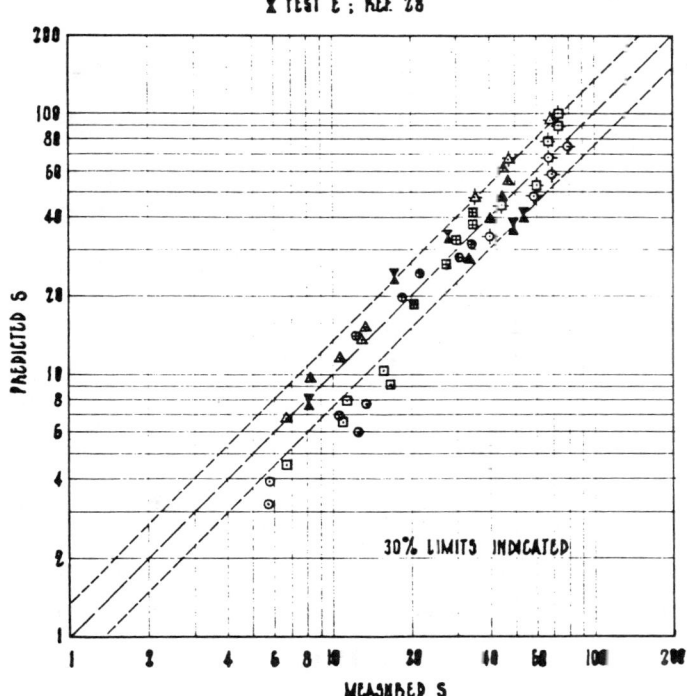

MODEL CALIBRATION AND VERIFICATION - S VALUES

FIGURE 43.

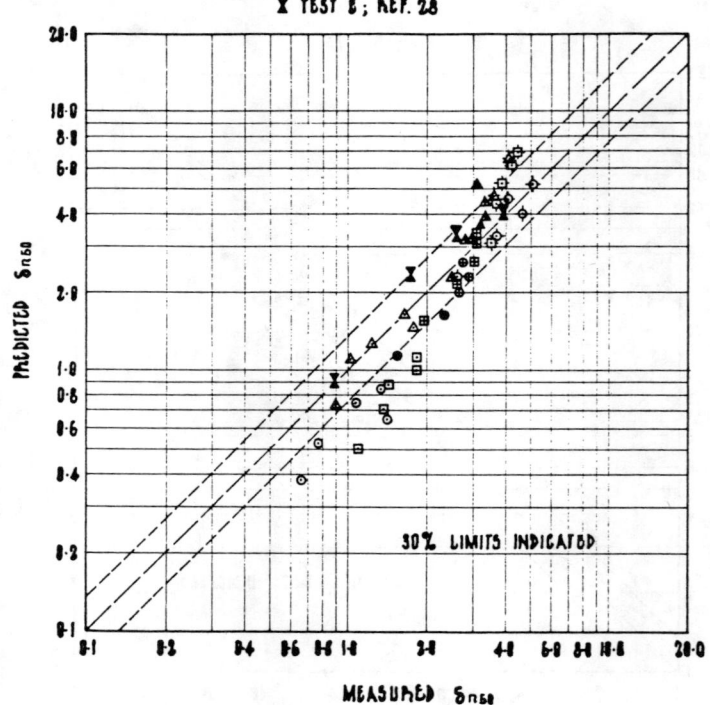

MODEL CALIBRATION AND VERIFICATION – S_{n50} VALUES

FIGURE 44.

/DESIGN_CASE
Tug Harbour Breakwater - Trunk 100 yr

/DENSITIES
    2.72   1.025

being ρ_a and ρ_w (tonnes/m^3) respectively.

/GEOMETRY
    1.35   18.   300.

being $Cot\alpha$, B_H (m) and B_L (m) respectively.

/STORM
100 yr Tropical Cyclone "Alpha"
5.0   5.0   8.0

being H_s (m), T_z (s) and t_d (hrs) respectively.

Because the economics of construction are also to be considered, estimates of the cost per tonne of placing armour on the structure is represented to the model as four pairs of tonne and $/tonne figures forming a table as follows:

/ARMOUR_COSTS
4
2.0   15.0
5.0   20.0
10.0   45.0
20.0   60.0

such that the model will interpolate as necessary between these values when calculating costs.

The preceding information is all that is required to run the model (with the exception of some additional job and user identification) for this particular design case.

If no specific type of output is requested, the model produces only the matrix of W_{50} values which would be "stable" under the design conditions, or rather, with a given thickness of N_{n50} layers would result in a given number of δ_{n50} eroded layers. The present output options are:

/OUTPUT_W50   - as just described (the default)
/OUTPUT_DN50 - the corresponding armour sizes
/OUTPUT_NS   - Hudson's Stability Number Matrix
/OUTPUT_KD   - Hudson's Damage Coefficient Matrix
/OUTPUT_MPA   - Armour mass per unit surface area of structure
/OUTPUT_MTOT - Total armour mass matrix
/OUTPUT_$PA   - Placed cost per unit surface area
/OUTPUT_$TOT - Total armour placed cost
/OUTPUT_$OPT - The least capital cost alternative selected from the above matrices.

Figure 45 shows the range in W_{50} to survive the 100yr storm to varying degrees of damage, given the specified geometry. For example, with an armour mass of 4 tonnes and N_{n50} around 9.0, the model indicates a δ_{n50} value of approximately 2.5 eroded layers for this option. If larger armour units are used, but with the same number of layers then the damage is reduced. Alternatively, the damage is reduced by increasing the number of layers for a given armour mass. The contours of W_{50} indicate where the stability changes most rapidly. In this example, it becomes difficult to limit the number of eroded layers to less than one without incurring increasingly more severe unit mass penalty. Between two and five eroded layers, the variation is much more gradual and smaller increases in unit mass yield reasonable decreases in eroded depths. Areas of the matrix shown as "-" indicate technically the failure region, here taken to be less than two layers remaining. Areas shown as "+" indicate the armour mass was either in excess of the upper mass limit to be considered (here 20 tonnes) or was below the lower limit (0.1 tonnes).

Note that the model has issued warnings to the effect that both the values of Cotα and Meer's P (from N_{n50}) which have been used in forming the matrix, are outside the experimental ranges so far tested and therefore further caution should be used in interpretation.

Figure 46 shows the corresponding matrix of armour cost per unit surface area by also considering the cost per tonne of the particular W_{50} value for each option. The solution surface in this case shows there are some cost advantages as a function of the number of layers placed if two eroded layers can be tolerated, i.e. from \$4.1/m^2 to \$6.3/m^2. With a different cost structure, the result could have been more pronounced and may have clearly indicated a particular course of action. In any case, there may be other reasons for selecting a particular option, e.g. the costs for 3.5 layers with 1.0 eroded layers are comparable to the costs for 11.0 layers with 1.5 eroded. Overall construction times for the two options could however be quite different with one more prone to disruptions by weather or industrial dispute etc.

Figure 47 shows the model's selection of the lowest capital cost design for the given design ranges. In this case it is 6.0 layers of 3.3 tonne armour which could be eroded a further 4.0 layers after the 100 year return period storm. The model does not yet balance this capital cost against the likely maintenance costs in restoring the structure to its former level of stability, but this could be included based on design estimates.

The functional form of the empirical Meer model has been shown to extend reasonably well for the case of the highly permeable (multi-layered) design. Through a systematic approach to model testing, a relationship has been proposed which relates the Meer permeability coefficient to the relative thickness of armour layers overlaying an impermeable core. Several practical deficiencies in the application of the Meer model have been overcome or improved by reformulation.

ARMOURED BREAKWATER

```
[HARBREM V1.0] [Example Breakwater Design    ] [B.A.Harper  ] [In:example.inp ] [Out:EXAMPLE.OUT ] [/OUTPUT_W50 ] [ 7-MAY-87 12:39]
/DESIGN_CASE    Tug Harbour Breakwater - Trunk 100yr              [WARNING: COTalpha beyond experimental range]
                                                                   [WARNING: Meer's P beyond experimental range]
/STORM          100yr Tropical Cyclone "Alpha"
                Hs = 5.00 m   Tz = 5.00 s   td = 8.00 hr

/DENSITIES      ROHa = 2.720 tonnes/m**3   ROHw = 1.025 tonnes/m**3    [ Surface Area    = .91E+04 m*m    ]

/GEOMETRY       COTalpha = 1.35   BH = 18.00 m   BL = 300. m           [ Surf Similarity = 2.07 (Plunging) ]
```

	MINIMUM ARMOUR MASS W50 (tonnes)																	
Deln50 Max. No. Eroded Layers	Nn50 - No. Layers of Armour																
	2.0	2.5	3.0	3.5	4.0	4.5	5.0	5.5	6.0	6.5	7.0	7.5	8.0	8.5	9.0	9.5	10.0	10.5	11.0
0.2	-	+	+	+	+	+	+	+	+	+	+	+	+	+	+	+	+	+	+
0.4	-	+	+	+	+	+	+	19.3	18.5	17.8	17.2	16.6	16.1	15.6	15.1	14.7	14.3	13.9	13.6
0.6	-	-	18.3	17.2	16.3	15.6	14.9	14.2	13.7	13.2	12.7	12.3	11.9	11.5	11.2	10.9	10.6	10.3	10.0
0.8	-	-	14.7	13.9	13.2	12.5	12.0	11.5	11.0	10.6	10.2	9.9	9.6	9.3	9.0	8.7	8.5	8.3	8.1
1.0	-	-	12.5	11.8	11.1	10.6	10.1	9.7	9.3	9.0	8.7	8.4	8.1	7.8	7.6	7.4	7.2	7.0	6.8
1.5	-	-	-	8.7	8.2	7.8	7.5	7.2	6.9	6.6	6.4	6.2	6.0	5.8	5.6	5.5	5.3	5.2	5.0
2.0	-	-	-	-	6.6	6.3	6.0	5.8	5.5	5.3	5.1	5.0	4.8	4.7	4.5	4.4	4.3	4.2	4.1
2.5	-	-	-	-	-	5.3	5.1	4.9	4.7	4.5	4.4	4.2	4.1	3.9	3.8	3.7	3.6	3.5	3.4
3.0	-	-	-	-	-	-	4.4	4.3	4.1	3.9	3.8	3.7	3.5	3.4	3.3	3.2	3.2	3.1	3.0
3.5	-	-	-	-	-	-	-	3.8	3.6	3.5	3.4	3.3	3.2	3.1	3.0	2.9	2.8	2.7	2.7
4.0	-	-	-	-	-	-	-	-	3.3	3.2	3.1	3.0	2.9	2.8	2.7	2.6	2.5	2.5	2.4
5.0	-	-	-	-	-	-	-	-	-	2.6	2.5	2.4	2.3	2.3	2.2	2.2	2.2	2.1	2.0

FIGURE 45.

```
[HARBREM V1.0] [Example Breakwater Design  ] [B.A.Harper  ] [In:example.inp ] [Out:EXAMPLE.OUT ] [/OUTPUT_SPA ] [ 7-MAY-87 12:39]
/DESIGN_CASE    Tug Harbour Breakwater -  Trunk 100 yr         [WARNING: COTalpha beyond experimental range]
                                                                [WARNING: Meer's P beyond experimental range]
/STORM          100yr Tropical Cyclone "Alpha"
                Hs = 5.00 m   Tz = 5.00 s   td = 8.00 hr
/DENSITIES      RUMa = 2.720 tonnes/m**3   RUMw = 1.025 tonnes/m**3   [ Surface Area    = .91E+04 m*m    ]
/GEOMETRY       COTalpha = 1.35   BH = 18.00 m   BL = 300. m          [ Surf Similarity = 2.07 (plunging) ]
```

ARMOUR COST PER UNIT SURFACE AREA ($100's / m*m)

Dein50 Max. No. Eroded Layers	2.0	2.5	3.0	3.5	4.0	4.5	5.0	5.5	6.0	6.5	7.0	7.5	8.0	8.5	9.0	9.5	10.0	10.5	11.0
0.2	-	+	+	+	+	+	+	+	+	+	+	+	+	+	+	+	+	+	+
0.4	-	+	+	+	+	+	+	10.2	10.7	11.3	11.8	12.3	12.8	13.3	13.7	14.2	14.6	15.0	15.5
0.6	-	-	5.3	5.9	6.5	7.0	7.5	8.0	8.5	8.9	9.4	9.8	10.2	10.6	11.0	11.4	11.8	12.1	12.5
0.8	-	-	4.5	5.0	5.5	6.0	6.4	6.8	7.3	7.7	8.1	8.4	8.5	8.6	8.8	8.9	9.0	9.1	9.1
1.0	-	-	4.0	4.4	4.9	5.3	5.7	6.0	6.1	6.3	6.4	6.5	6.7	6.8	6.8	6.9	7.0	7.1	7.1
1.5	-	-	-	3.2	3.4	3.6	3.7	3.8	3.9	4.0	4.1	4.2	4.2	4.3	4.3	4.4	4.4	4.4	4.5
2.0	-	-	-	-	2.5	2.6	2.7	2.8	2.8	2.9	2.9	3.0	3.1	3.2	3.3	3.5	3.6	3.7	3.8
2.5	-	-	-	-	-	2.0	2.1	2.2	2.3	2.4	2.5	2.6	2.8	2.9	3.0	3.1	3.2	3.3	3.4
3.0	-	-	-	-	-	-	1.8	2.0	2.1	2.2	2.3	2.4	2.5	2.6	2.7	2.8	2.9	3.0	3.1
3.5	-	-	-	-	-	-	-	1.8	1.9	2.0	2.1	2.2	2.3	2.4	2.5	2.6	2.7	2.8	2.9
4.0	-	-	-	-	-	-	-	-	1.8	1.9	2.0	2.1	2.2	2.3	2.4	2.5	2.5	2.6	2.7
5.0	-	-	-	-	-	-	-	-	-	-	1.8	1.9	2.0	2.1	2.1	2.2	2.3	2.4	2.5

FIGURE 46

ARMOURED BREAKWATER

```
[HARBREM V1.0] [Example Breakwater Design   ] [B.A.Harper  ] [In:example.inp ] [Out:EXAMPLE.OUT ] [/OUTPUT_SOFT] [ 7-MAY-87 12:39]

/DESIGN_CASE    Tug Harbour Breakwater -  Trunk 100yr           [WARNING: COTalpha beyond experimental range]
/STORM          100yr Tropical Cyclone "Alpha"                  [WARNING: Meer's P beyond experimental range]
                Hs = 5.00 m   Tz = 5.00 s   td = 8.00 hr

/DENSITIES      ROHa = 2.720 tonnes/m**3   ROHw = 1.025 tonnes/m**3    [ Surface Area   = .91E+04 m*m     ]

/GEOMETRY       COTalpha = 1.35   BH = 18.00 m   BL = 300. m           [ Surf Similarity = 2.07 (plunging) ]
```

```
: MINIMUM CAPITAL COST :
: ARMOUR DESIGN        :

Unit Armour Mass :      W50    = 3.3 tonnes

Nom. Armour Dia. :      Dn50   = 1.1 m

Nom. Armour Vol. :      Vn50   = 1.2 m*m*m

Unit Armour Cost :      Cn50   = $ 56.54

Placed Thickness :      Nn50   = 6.0 layers

Erodable Thickness:     Deln50 = 4.0 layers


Per Unit Surface Area of Armour Slope:

Mass   =   10.44 tonnes/m*m

Volume =   3.84 m

Cost   = $ 179.10 /m*m


TOTALS:

Mass   =   .95E+05 tonnes

Volume =   .35E+05 m*m*m

Cost   = M$ 1.62
```

FIGURE 47

A FORTRAN 77 computer program has been developed which not only simplifies the design process, but provides new insights to the structure behaviour and assists in optimising capital costs. Overall accuracy of the model developed here is of the order of 30 % when compared with that data, which is a good result for the prediction of rubble mound breakwater stability. The model should be used as an initial design tool in selecting appropriate armour sizes for physical model testing only, to ensure other possible effects (e.g. 3-D) are adequately addressed by the designer.

CONCLUSION

This project introduced into Australia the potential for using hydraulic models to maximise the output of quarried materials in phase with the construction of rubble mound breakwaters. This can result in very significant savings in cost and the minimising of risk to the principal and the contractor. In this way the efficiency of all parties to the project may be markedly increased.

For contractors, weather factors and rock materials source/s in the construction of rubble mound breakwaters, usually represent very high risk ventures. Where risk reduction is achieved, price reduction is its natural corollary. In this contract a deliberate attempt was made in the documentation and subsequent construction supervision to minimise construction risks.

This project demonstrated that when a knowledgeable principal is allied with competent engineering and an experienced contractor, a satisfactory completion of a high risk project is the end result.

ACKNOWLEDGEMENT

The Authors acknowledge with thanks, the co-operation and assistance given to this project by the Principals - "DBCT - UDCL Joint Venture"; Department of Harbours and Marine, Queensland; University of New South Wales Hydraulics Research Laboratory and the Contractors - Roche Bros. Australia Pty Ltd.

REFERENCES

AHRENS J.P. & McCARTNEY B.L.
"Wave Period Effect on the Stability of Riprap"
Proc Civil Eng in the Oceans III, Vol 2, 1975, pp 1019-1034

BATTJES J.A.
"Surf Similarity"
Proc 14th I.C.C.E., ASCE, Copenhagen 1974, pp 466-479

BLAIN BREMNER & WILLIAMS PTY LTD
Half Tide Tug Harbour - Extreme Water Level Study
DBCT - UDCL Joint Venture, September 1983

BLAIN BREMNER & WILLIAMS PTY LTD
Extreme Water Level Study
Addendum: Extreme Wave Height Frequencies
DBCT - UDCL Joint Venture, January 1984

BLAIN BREMNER & WILLIAMS PTY LTD
Hydraulic Model Testing of Proposed Re-Designs of the Breakwater
DBCT - UDCL Joint Venture, Half Tide Tug Harbour, March 1986

BREMNER W., FOSTER D.N., MILLER C.W., WALLACE B.C.
"The Design Concept of Dual Breakwaters and its Application to Townsville, Australia", Proc 17th I.C.C.E., Sydney, 1980, pp 1898-1908

FOSTER D.N., McGRATH B.L., BREMNER W.
"Rosslyn Bay Breakwater Queensland Australia"
Proc 16th I.C.C.E., Hamburg 1978

FOSTER D.N., HARADASA D.K.C.
"Rosslyn Bay Boat Harbour Breakwater Model Studies"
Tech Rep 77/06. The University of New South Wales, Water Research Laboratory.

FOSTER D.N., MILLER C.A., WALLACE B.C.
Townsville Harbour Eastern Breakwater Extension Hydraulic Model Studies. Tech Rep 80/01. The University of New South Wales, Water Research Laboratory, January 1980.

FOSTER D.N., McGRATH B.L., BREMNER W.
"Half Tide Tug Harbour Hydraulic Model Studies", The University of New South Wales, Water Research Laboratory, Tech Rep 83/15, January 1984.

FOSTER D.N., COX R.J., HILLS J.E.
"Half Tide Tug Harbour - Hydraulic Model Testing of the Proposed Re-Design of the Breakwater", The University of New South Wales, Water Research Laboratory, Tech Rep 86/02, May 1986.

FOSTER D.N., HARADASA D.K.C., FOSTER S.J.
Half Tide Tug Harbour Hydraulic Model Studies. The University of New South Wales, Water Research Laboratory, Tech Rep 83/15, January 1984.

GODA Y.
"A Synthesis of Breaker Indices". Proc Japan Soc Civil Engineers No. 180, 1970.

HARBOURS CORPORATION OF QUEENSLAND
Hay Point Tug Harbour Contract Documents No. HP3 - Breakwater Construction, November 1976.

HARPER B.A.
"Half Tide Tug Harbour Tropical Cyclone Spectral Wave Modelling Study", prepared for DBCT-UDCL Joint Venture, Blain Bremner & Williams Pty Ltd, September 1983.

HARPER B.A.
"Stability of Highly Permeable Breakwaters", prepared under Australian Marine Sciences and Technologies Grants Scheme 1985, Blain Bremner & Williams Pty Ltd, April 1987.

HIGGS K, FOSTER D.N.
Model Testing of the Eastern Breakwater Hay Point Tug Harbour. The University of New South Wales, Water Research Laboratory, Tech Rep 77/12, September 1977.

HUDSON R.Y., DAVIDSON D.D.
"Reliability of Rubble Mound Breakwater Stability Models", 2nd ASCE Sym Modelling Techniques 1975.

LOSADA M.A., GIMENEZ-CURTO L.A.
"Joint Effect of the Wave Height and Period on the Stability of Rubble Mound Breakwaters using Iribarren's Number".
Coastal Eng v 3 n 2 Dec 1979, pp 77-96.

LOSADA M.A., GIMENEZ-CURTO L.A.
"Flow Characteristics on Rough Permeable Slopes Under Wave Action" Coastal Eng, 4, 1981, pp 187-206.

MEER J.W. VAN DER, PILARCZYK K.W.
"Stability of Rubble Mound Slopes Under Random Wave Attack", Proc 19th I.C.C.E., ASCE, Houston, September 1984.

MEER J.W. VAN DER
"Stability of Rubble Mound Revetments and Breakwaters Under Random Wave Attack", Proc Breakwaters '85 Conf, London, October 1985.

QUEENSLAND GOVERNMENT HYDRAULICS LABORATORY
Hay Point Tug Harbour Structural Model of an Overtopped Breakwater. Report No. M11/1 June 1981.

SOBEY R.J., YOUNG I.R.
"Hurricane Wind Waves - A Discrete Spectral Model"
ASCE Jnl Waterway Port Coastal & Ocean Engineering, Vol 112, No. 3, May 1986, pp 370-389.

THOMPSON D.M., SHUTTLER R.M.
"Riprap Design for Wind Wave Attack - A Laboratory Study in Random Waves".
Hydraulics Research Station, Wallingford, EX 707, September 1975.

U.S. ARMY CORPS OF ENGINEERS
"Shore Protection Manual", Coastal Engineering Research Centre, 1977.

PERFORMANCE OF A BERM ROUNDHEAD IN THE ST. GEORGE BREAKWATER SYSTEM

by

Jeffrey F. Gilman

Abstract

A new harbor under construction on St. George Island in Alaska's Bering Sea is using the berm breakwater concept for protection from wave attack. Three breakwaters are included in the system, two outer breakwaters to protect an entrance channel, and an inner breakwater to protect an 8-acre moorage basin. In late 1986 the designers were faced with a shutdown in construction with the North Breakwater roundhead only half finished. There was a question as to the capacity of the structure to withstand wave attack. The purpose of this paper is to demonstrate that during the winter of 1986-87 storms occurred which approached the design storm in intensity and that, even with the half-complete nature of the structure, the berm roundhead performed very well and suffered only minor berm profile modification.

Résumé

Un nouveau port en construction sur l'île St. George dans la mer de Béring en Alaska sera protégé des vagues par des brise-lames de type à risberme. Le système prévoit trois brise-lames, deux brise-lames extérieurs protégeant un chenal d'entrée et un brise-lames intérieur abritant un bassin d'amarrage de 8 acres. Vers la fin de 1986 les concepteurs ont vu les travaux interrompus alors que le musoir du brise-lames nord n'était qu'à demi complété. On s'interrogeait quant à la possibilité que l'ouvrage résiste à l'assaut des vagues. Le but de la présente étude est de démontrer que même si des tempêtes d'une intensité approchant celle de la tempête nominale se sont abattues sur l'ouvrage à demi achevé pendant l'hiver de 1986-1987, le musoir à risberme s'est très bien comporté et qu'il n'y a eu qu'une modification mineure du profil de la risberme.

PERFORMANCE OF A BERM ROUNDHEAD IN THE ST. GEORGE BREAKWATER SYSTEM

Jeffrey F. Gilman,* A.M. ASCE

INTRODUCTION

A new harbor under construction on St. George Island in Alaska's Bering Sea is using the berm breakwater concept for protection from wave attack (Orage and Gilman, 1987). Three breakwaters are included in the system, two outer breakwaters to protect an entrance channel, and an inner breakwater to protect an 8-acre moorage basin as shown in Fig. 1. The typical section for the outer breakwaters' roundheads is shown in Fig. 2.

In late 1986 the designers were faced with a shutdown in construction with the North Breakwater roundhead only half finished. There was a question as to the capacity of the structure to withstand wave attack. The purpose of this paper is to demonstrate that during the winter of 1986-87 storms occurred which approached the design storm in intensity and that, even with the half-complete nature of the structure, the berm roundhead performed very well and suffered only minor berm profile modification.

DESIGN STORM PARAMETERS

Typical sections for the breakwaters were developed in extensive physical model testing performed primarily at the Danish Hydraulic Institute (1983) and the Delft Hydraulics Laboratory (1985). The final sections were developed at Delft using a "worst-case" storm. This was possible due to the depth limitations on wave height growth. Based on a visual inspection in a flume the waves in the six-step storm series comprising each 36-hour test of the structure produced the greatest number of waves breaking directly on the structure.

---

* Senior Engineer, Peratrovich, Nottingham & Drage, Inc., 1506 W. 36th, Anchorage, Alaska, USA 99503

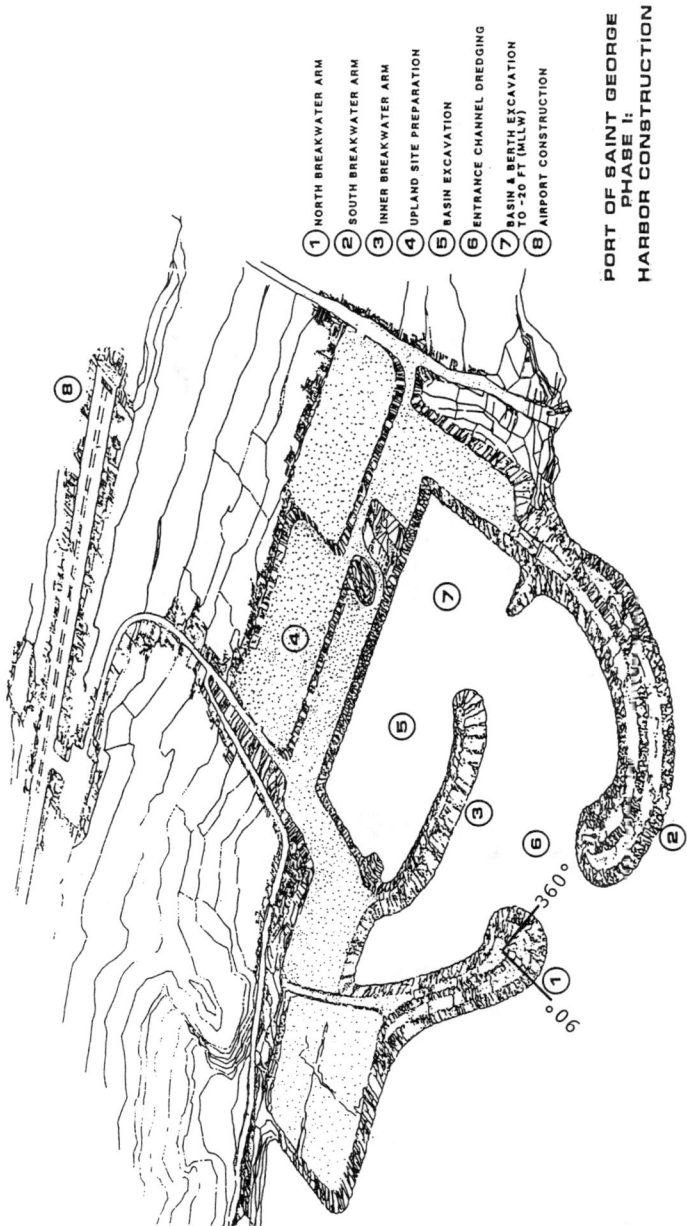

FIGURE 1

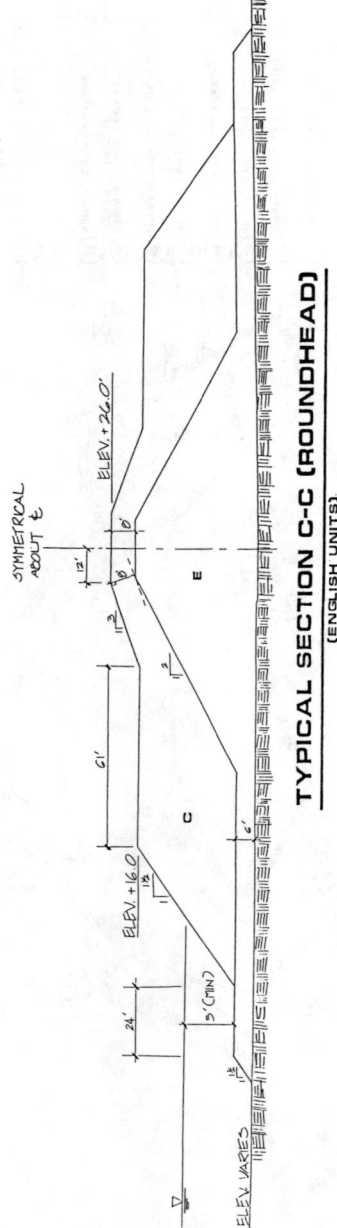

TYPICAL SECTION C-C (ROUNDHEAD)
(ENGLISH UNITS)

FIGURE 2

TABLE 1

FINAL STORM PROFILE FOR 1985 DELFT PHYSICAL MODEL

Step	Hs (m)	Tp (s)
1	6.8	8.7
2	8.3	11.4
3	8.9	14.3
4	10.0	15.9
5	11.2	16.9
6	7.2	23.8

PROTOTYPE STORMS

Because of the depth-limiting characteristics we expect portions or all of the design storm to occur at least once each year during the period of frequent storm activity, October through March. The worst storms for each of the last three years at the site from the author's subjective point of view are listed below.

Date	Duration	Maximum Deepwater Hs	Maximum Tp
7-8 Dec. 1984*	35 hrs.	11.0 m.	15 secs.
30 Nov. - 2 Dec. 1985**	46	7.6	17
30 Nov. - 2 Dec. 1986**	60	8.8	17

*Waverider Wave Buoy located approx. 100 km. southeast of St. George
** Waverider located approx. 300 km west of St. George

A comparison of this table with the design storm steps presented in Table 1 indicates that Step 4 of the design storm was nearly equaled in the 1985 and 1986 storms while Step 5 coincides with the peak of the 1984 storm (which it was consciously intended to resemble, the 1984 storm data being available at the time of modeling at Delft in Spring 1985.) A further examination of Bering Sea Wave Buoy records for the winter 1986-87 indicates an additional four storms, the peaks of which fall within the realm of "moderate" or "severe" storm, that is, those storms with waves exceeding 4.6 m. in height (thus capable of producing maximum wave conditions of approximately 6 m. at the breakwaters after shoaling). These storms are summarized below. Note that while the storm of late January-early February seems slightly more severe than the late November-early December storm listed above, the listed parameters were recorded at some distance from St. George while the corresponding conditions at the site were more severe in the case of the late November-early December storm than the later storm.

Date	Duration	Maximum Deepwater Hs	Maximum Tp
27-29 Nov.	46 hrs.	8.5 m.	14 s.
3-4 Dec.	24	6.1	13
13-14 Jan.	25	7.0	11
26 Jan. - 1 Feb.	137	9.1	17

NORTH BREAKWATER ROUNDHEAD

When construction ended on the North Breakwater in the Fall of 1986, the North Breakwater was only partially complete. The berm portion of the trunk - that part seaward of centerline to elev. +3.7 m. (MLLW) was virtually complete, while the berm portion of the roundhead was only about half complete, with as little as 9.1 m. of horizontal berm none of which rose above elev. +3.7 m. (MLLW). The design called for +4.9 m. (MLLW). The gradation of rock in the roundhead generally met specification (2-10 tons), with some contamination (undersize material which plugs the voids necessary for proper berm performance) resulting from the contractor's crane pads primarily on the rear of the roundhead.

WINTER 1986-87 SURVEYS

Prior to the onset of moderate and severe storms in October the author surveyed the North Breakwater to determine a baseline for subsequent measurements of berm development. Lines composed of alkyd paint dots brushed on the top of stones were put out on the roundhead at bearings of 270°, 360°, and 90° from the center point of the roundhead as well as at cross-sections every 100 ft. along the trunk of the breakwater. The locations of two of these profiles, the 360° and 90° profiles, are shown in Fig. 1.

These paint marks were intended to be re-surveyed at intervals to determine the movement of the painted rocks, and thus the berm in general. Because of the highly porous nature of the berm, that is, because of alternating deep voids and high points in the structure, it was difficult to determine totally representative cross-sections. Therefore, the cross-sections presented are not necessarily totally accurate representations of the berm as a whole, only the specific sections profiled.

Surveys of the three roundhead profiles were performed on 24 Oct. 1986 and again on 29 Oct. immediately following two moderate storms which occurred in quick succession. On 20 Nov. profiles were surveyed just prior to the severe late Nov.-early Dec. storm. Then they were surveyed again on 22 Jan. 1987 just prior to the severe storm of late January, and on 27 Feb. following most of the moderate or severe storms of the winter.

DISCUSSION AND CONCLUSION

An examination of the profiles done on 24 Oct. and 29 Oct. shows very little change. Indeed, little significant change is noted until the last profiles done in Feb. 1987. Because of this lack of change in successive surveys and because the 29 Oct. survey was the most complete of the earlier surveys, it is shown in Figs. 3-4 along with the survey of 27 Feb. in order to gain an appreciation of the berm profile before and after the winter of 1986-87.

The storms of late Nov.-early Dec. 1986 and late Jan.-early Feb. 1987 were severe and came close to equaling Step 4 of the "worst case" storm used in the physical model. The effect of these two storms can be seen most clearly in Fig. 4 where some subsidence of the berm seems evident in the 90° profile ("right"). At points at the rear of the berm and about midway along the berm, the stones have subsided, or at least shifted, so that the berm is approximately 1 m. lower in those places. There is also a slight, but noticeable regression of the face of the berm. The North Roundhead is built on a layer of sand probably 3 m. thick on average. Although there is a 2 m. filter layer beneath the armor stone the designers still expect some subsidence of the berm will occur due to scour of the sand layer. It is probable that this accounts for some of the subsidence shown in Fig. 4. In addition, the physical models indicate that the berm consolidates under wave attack until it becomes a tightly nested mass. This is also probably occuring in the incomplete North Roundhead and may partially account for the subsidence, but more importantly is probably a major factor in the slight regression of the berm face shown in Fig. 4.

The profile for the 360° profile ("ahead") however, shows no subsidence and, fact, one stone has been thrust upward at the seaward edge of the berm, a feature typical of the physical model, as well, in certain cases. It should be noted that this section of the roundhead was subjected to wave attack as severe as that suffered by the section of the roundhead represented by the 90° profile. The South Breakwater, which will provide some protection, was not in place in the winter of 1986-87.

The data presented indicates that, even under the severe wave attack typical of the annual event at St. George, and in spite of

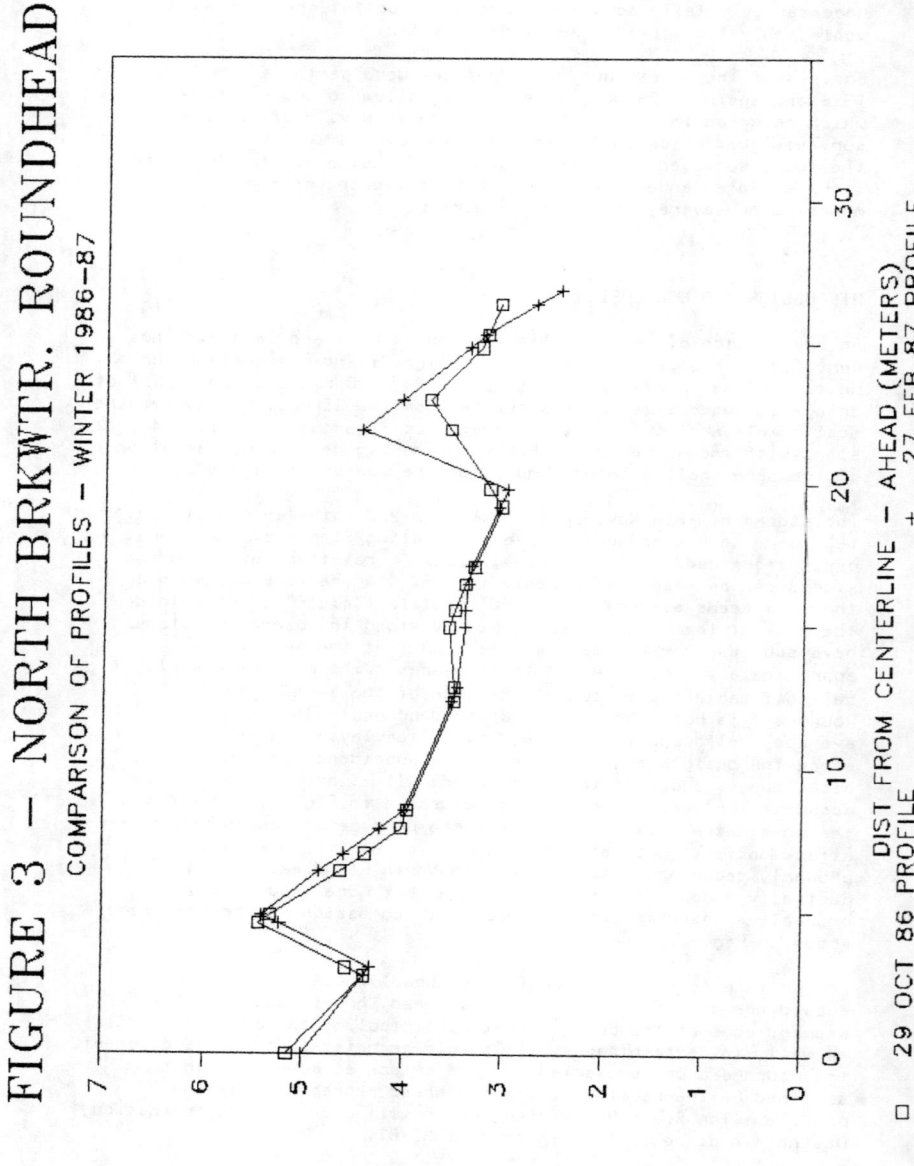

FIGURE 3 – NORTH BRKWTR. ROUNDHEAD
COMPARISON OF PROFILES – WINTER 1986-87

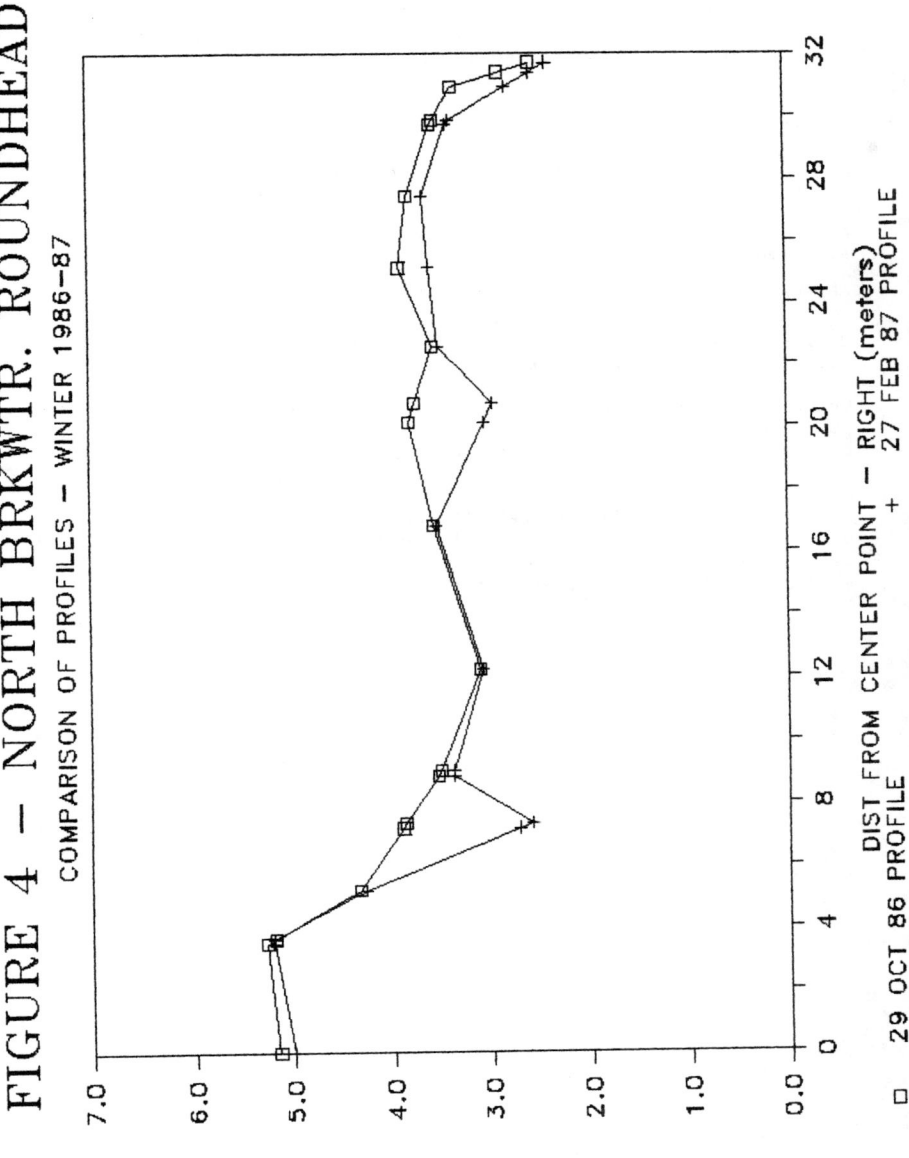

FIGURE 4 — NORTH BRKWTR. ROUNDHEAD
COMPARISON OF PROFILES — WINTER 1986-87

the half-complete nature of the breakwater in place last winter,
the St. Geroge berm roundhead holds up well and, in fact, seems to
perform better than physical modeling would indicate.

ACKNOWLEDGEMENTS

The author wishes to thank the City of St. George represented by
Mayor Max Malavansky and City Administrator Rich Wilson for their
open-minded approach as clients on this project, Brent Drage for
reviewing the draft of this paper, and Greg McGlashan for his
assistance in the surveys required.

REFERENCES

Danish Hydraulic Institue (1983). St. George Harbour, Pribilof
Islands, Alaska. hydraulic model investigation. Unpublished.

Delft Hydraulics Laboratory (1985). St. George Harbor, Alaska,
two-dimensional tests on scale effects, three-dimensional tests on
breakwater stability. Report on Model Investigation No. M2102.
Unpublished.

Drage, Brent T. and Gilman, Jeff (1987). St. George Harbor - a
berm breakwater solution in Alaska's Bering Sea. 1987 Proceedings
of Coastal & Port Engineering in Developing Countries, Beijing,
Nanjing Hydraulic Research Institute, Vol. 1, Pp. 550-556.

IMPLEMENTATION AND PERFORMANCE OF BERM BREAKWATER DESIGN AT RACINE, WI

by

Robert J. Montgomery,
Gregory J. Hofmeister and William F. Baird

Abstract

A large berm-type breakwater has been constructed at the entrance to Racine Harbor, Wisconsin, on the western shore of Lake Michigan. The breakwater is approximately 450 feet long, in water depths of 20 to 25 feet, and utilizes dolomite armour stone in the weight range 300 to 8000 lbs. The design storm maximum significant wave height was 14.5 feet. The berm design was developed in an extensive physical modeling study, with construction plans and contract specifications developed using the model study results and knowledge of local construction materials and contracting conditions. Construction of the breakwater was completed in the Fall of 1986. This paper focuses on the process of building the berm breakwater in a bid/contract environment, and on the performance of the breakwater in major storms which occured in February and March of 1987.

Résumé

Un gros brise-lames de type à risberme a été construit à l'entrée du port de Racine au Wisconsin sur le rivage ouest du lac Michigan. Il est d'une longueur approximative de 450 pieds, par des profondeurs d'eau de 20 à 25 pieds et construit de pierres de carapace de dolomie d'une masse variant de 300 à 800 lb. La hauteur significative maximale des vagues de la tempête nominale était de 14,5 pieds. La conception de la risberme a été élaborée lors d'une étude approfondie de modélisation physique dont les résultats, ainsi qu'une connaissance des matériaux de construction disponibles sur place et des conditions locales d'attribution des contrats, ont permis d'élaborer les plans de construction et les spécifications du contrat. La construction du brise-lames était terminée à l'automne de 1986. Cette étude porte principalement sur le processus de la construction du brise-lames à risberme dans le cadre d'un contrat accordé après présentation de soumissions et sur le rendement du brise-lames lors de tempêtes majeures en février et en mars 1987.

IMPLEMENTATION AND PERFORMANCE OF BERM BREAKWATER DESIGN AT RACINE, WI

Robert J. Montgomery[1]
Gregory J. Hofmeister[2]
William F. Baird[3]

ABSTRACT

A berm breakwater has been constructed at the entrance to Racine Harbor, Wisconsin, on the western shore of Lake Michigan. The breakwater is approximately 450 ft (135 m) long, in water depths of 20 to 25 ft (6.1 - 7.6 m) and utilizes dolomite armor stone in the weight range 300 to 8000 lbs (0.14 - 3.63 t). The design storm maximum significant wave height was 14.5 ft (4.4 m). The berm design was developed in a physical modeling study, using data from local quarry test blasts in establishing the armor stone gradation. Construction of the breakwater was completed in November 1986. This paper focuses on the process of building the berm breakwater in a bid/contract environment, and on the performance of the breakwater in major storms which occurred in February and March of 1987.

Development of bid and design documents for construction of a berm-type breakwater can be difficult for a location where a variety of contractors and potential material sources exist. Construction documents and the final design must be carefully arranged, in order to realize the potential cost savings of using a berm design. At Racine, bid documents specifying a single design were utilized. Other options exist, including the use of alternate bid sections, which could be evaluated before the design process is underway. The paper describes bid, specification and construction quality control options that were used at Racine, and alternatives which could be useful on other projects.

Soon after the berm breakwater was complete, two major storms occurred on Lake Michigan, in February and March of 1987. The March storm produced wave height and storm duration conditions that approximated the design conditions used in the model study, while water levels exceeded design conditions. The berm breakwater underwent major readjustment in response to these storms. Surveyed cross sections indicate that the prototype response was closely simulated by the model study. However, some significant erosion near the water level occurred and was compensated for by placement of a small amount of quarrystone armor.

1, 2: Warzyn Engineering, P.O. Box 5385, Madison, Wisconsin 53705
3: W F Baird & Associates, 38 Antares Drive, Ottawa, Ontario K2E 7V2 CANADA

INTRODUCTION

The design which was developed for the Racine harbor entrance breakwater is of a class which has been referred to as "berm-type" or "unconventional" rubblemound breakwaters. The term "berm breakwater" is used in this paper. The major characteristics of the berm design are that 1) it is often built wiht a large mass of armor stones arranged in a berm, or shelf, on the exposed face of the breakwater, and that 2) this berm is expected to reshape into a quasi-stable profile under wave action. The berm breakwater derives its stability from the deep section of interlocking armor stone, which dissipates wave energy within the section. One advantage of a berm design is that the individual armor stone may be smaller and of a broader gradation than that required in more conventional 2-layer armor designs, which can result in considerable cost savings in certain projects.

This paper focuses on the implementation of the berm design developed for Racine Harbor and emphasizes bidding, specifications, construction management and breakwater performance in the storm conditions of March, 1987. The paper concludes with a summary of lessons learned from this Great Lakes application of berm breakwater construction.

PROJECT DESCRIPTION

The Racine Harbor entrance breakwater is located at the end of the old north federal breakwater at the City of Racine Harbor (Figure 1) located on the west shore of Lake Michigan, approximately 30 miles (19 km) south of Milwaukee, Wisconsin. The entrance breakwater was built as part of an extensive harbor and waterfront recevelopment plan conducted by the Racine Harbor Development Commission. This plan also included construction of harbor breakwater protection in addition to the entrance breakwater, dredging and filling, construction of a 920-slip marina, and considerable land-based recreational and commercial development. The overall project has been described in several articles and conference proceedings (Ziolkowski and Ryan, 1987).

The harbor entrance breakwater design, as well as that of several other of the breakwater structures, was developed in a series of 2-D and 3-D model tests conducted primarily by W.F. Baird and & Associates, Ottawa, under contract to Warzyn Engineering, Madison, Wisconsin. Plans and specifications for the breakwater were developed by Warzyn, and construction management was controlled by Racine County, with some participation by Warzyn Engineering and W. F. Baird. Construction of the entrance breakwater was completed in November of 1986 by Dunbar and Sullivan, Cleveland, Ohio, contractors. Severe storms struck the breakwater in February and March of 1987. The February storm produced major shoreline flooding in the Chicago area, but the March storm was much more severe at Racine.

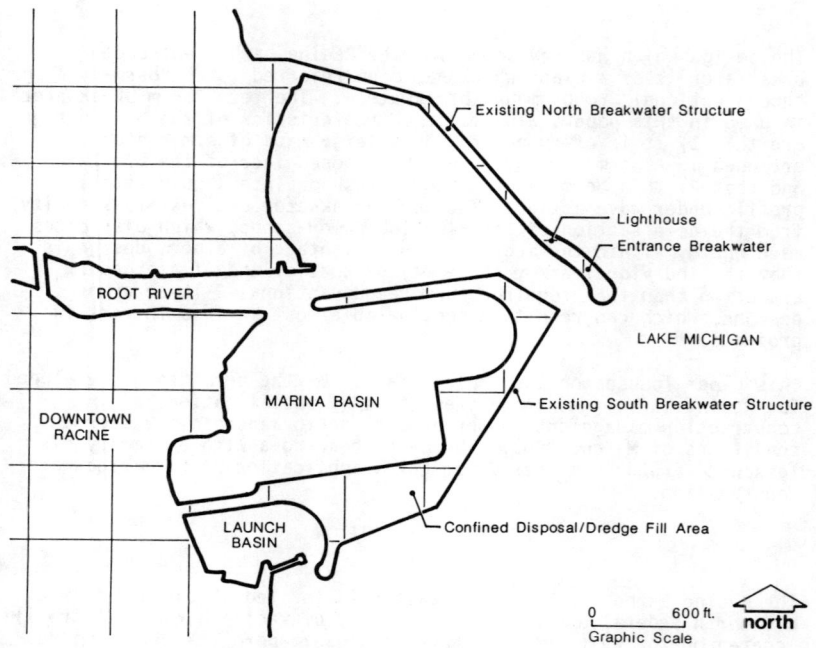

FIGURE 1 • RACINE HARBOR PROJECT LAYOUT

PHYSICAL SETTING

The Racine entrance breakwater is located on the exposed western shoreline of Lake Michigan. At this location, Lake Michigan is approximately 70 miles (110 km) wide, and the longest fetch length trends to the north-northeast, having a length of approximately 250 miles (400 km). Off-shore design wave conditions have been investigated in several studies, notably the hindcast/frequency analysis conducted by Resio and Vincent (1976). Deep water fall/winter storm wave characteristics at the 20-year return period were predicted to be highest from the northeast, with H_s= 18.3 ft (5.6 m) , T_s= 10.0 s, ranging down to H_s= 13.4 ft (4.1 m), T_s= 7.8 s from the southeast.

Water depth at the entrance to Racine Harbor is in the range of 20-25 ft (6.1 - 7.6 m). Wave approach from the southeast is complicated by the presence of the Racine Reef, while wave approach from the east and northeast encounters more regular bathymetry. Wave height at the harbor entrance was analyzed considering the effect of offshore bathymetry on shoaling and refraction by the Corps of Engineers. This study concluded that the highest 20-year storm waves approach from the northeast and have significant wave height at the harbor entrance of 14.6 ft (4.5 m).

Lake levels on the Great Lakes have recently reached all time highs. Maximum monthly levels on Lake Michigan have approached +5 ft, (1.5 m), with respect to low water datum, (LWD), more than 3 ft (0.9 m) above long-term average levels. In addition, static lake levels are increased by set-up during major storms. These high water levels have compounded shore protection problems throughout the Great Lakes, and have also cast doubt on existing statistical analyses of extreme lake levels. The Corps defined the 20-year boating season high water elevation as +4.5 ft, (1.4 m) LWD, including storm set-up.

DESIGN DEVELOPMENT

A berm-type breakwater design was initially suggested for use in augmenting the Racine north federal breakwater, to reduce overtopping of that structure. This design was refined and later constructed, but was somewhat overshadowed by the development and construction of the harbor entrance berm breakwater design. The preliminary project feasibility report (Warzyn, 1984) originally considered a sunken ore freighter for the main element in the harbor entrance breakwater. Later, the freighter breakwater proved technically infeasible, and a berm design entrance breakwater appeared much more economical than 2-layer armor stone conventional designs in preliminary cost comparisons.

Prior to commissioning final design modeling for the harbor entrance berm breakwater, a search of armor stone sources was conducted, to establish the gradation of material to be used in the model study. This search was influenced by the assumption that a local stone source would be the most economical, and centered around existing and potential quarries in the Racine area. The Vulcan Materials Company Ives quarry, located four miles from the harbor, contained a limited amount of Racine and Waukesha dolomite, which was concluded to be the best local prospect for providing both the quantity and quality of stone necessary for the project.

The Vulcan quarry was further investigated by performing several test blasts in the Racine dolomite unit, to develop gradation curves for armor and core stone production. Conceptually, an advantage of using material from a local stone source is to utilize the entire production of the quarry operation. Test blasting indicated that the yield of armor stone in normal operation should be in the 40-50% range, assuming a lower limit on armor stone of 300 lbs (0.14 t). This materials search indicated that the quantity of material required for the Racine project would not exceed the reserves of the quarry, nor produce large volumes of "waste" materials to be disposed of at additional cost to the quarry operation. It is important to note that final physical modeling and final design proceeded with the characterization of the local stone provided by this material search, and the implicit assumption that this local quarry would continue to be the cheapest supply source in an open bidding process.

Final design modeling for the entrance breakwater, as well as for other elements of the project, were conducted at the National Research Council of Canada Hydraulics Laboratories, Ottawa. Basic wave climate and water level design conditions were taken from the Corps report on Racine harbor (1978a) and on Lake Michigan high water levels (1978b), summarized in Table 1. The maximum significant wave lengths and periods from the Corps report were arranged into design storms, with peak intensities of approximately 10 h duration and total storm duration of 31 h. These storms were applied to various trial breakwater structures, using an irregular wave spectra generated in the 50-ft wide flume at the NRC laboratory. Stone for the model tests was selected from crushed limestone to meet the gradation from test blasts at the Vulcan Materials Racine quarry, with a W_{50}= 1800 lb (0.82 t) and D_{50}= 2.2 ft (0.68 m). Modeling was performed using Froude number similitude, at an undistorted geometric scale of 1:25. The overall conduct and results of final design modeling is described by Hall and Anglin (1985).

The entrance breakwater design process moved through several iterations, with the key decision variables being armor stone erosion and wave penetration through the harbor entrance. The final design configuration was slightly curvelinear in plan, and proved stable under design storm conditions from the northeast, east and southeast. The northeast design storm reduced the modeled width of the berm substantially (from 40 ft (12.2 m) to 10 ft (3.0 m)) in reshaping, but the berm remained stable thereafter. Armor stone movement was initiated at a wave height of 11 ft (3.35 m). The final design plan and cross section are shown in Figures 2 and 3 and summarized in Table 1.

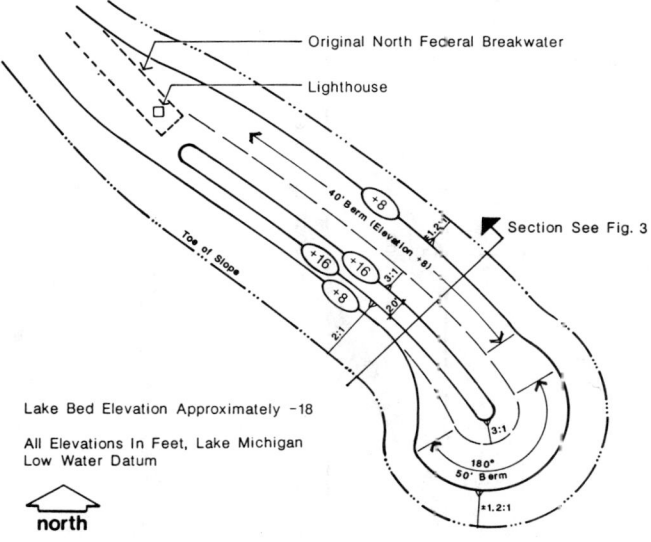

FIGURE 2 • PLAN VIEW – ENTRANCE BREAKWATER

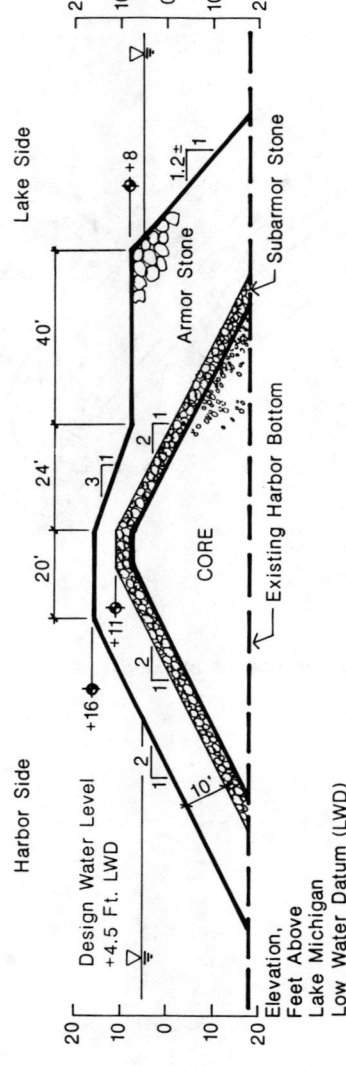

FIGURE 3 • TYPICAL SECTION THROUGH ENTRANCE BREAKWATER

Modeled Wave and Water Level

Highest design storm wave height: (From Northeast)	H_s = 14.5 ft (4.42 m) T_s = 10.0 sec
Design water level:	+4.5 ft, LWD
Water depth at structure:	27.5 ft (8.39 m)

Stone Materials

Silurian-age dolomite armor:	W_{50} = 1800 lb (0.82 t) D_{50} = 2.2 ft (0.68 m)

Breakwater Dimensions

Crest Length	450 ft (137 m)
Crest elevation	+16 ft, LWD
Berm elevation	+8 ft, LWD
Overall width at design waterline	approx 110 ft (33.5 m)
Armor (300 to 8,000 lbs dolomite)	70,000 tons (63,500 t)
Core (0-300 lbs dolomite)	52,000 tons (47,200 t)

Construction

Unit price contract paid on weight in verified sections.
100% land-based equipment.

Core: Dump truck, dozed into section.

Armor: Dumped, then dozed or placed with large backhoe.

Construction Period: April through November, 1986

Cost:

Armor:	US$ 1,260,000
Core:	US$ 468,000

Note: Refer to Figures 2 and 3 for plan view and cross-section of the breakwater.

Table 1. Vital Statistics: Racine Harbor Entrance Berm Breakwater

A dimensionless relative wave height parameter has been proposed for characterizing coastal structures, from stable rubblemound breakwaters through sand beaches (van der Meer and Pilarczyk, 1986). This parameter, given as $H_s/\Delta D_{n50}$ (see appendix) is 4.07, which falls in the range defined as "berm breakwaters and S-shaped profiles".

PLANS, SPECIFICATIONS AND BIDDING

Given the modeling results, it was a generally straightforward task to incorporate the alignments, cross sections and gradations into the design to produce a set of breakwater plans and specifications. Several areas which were not defined in detail by the modeling included connections of typical sections which had been modeled, such as the connection of the landward end of the entrance breakwater with the existing lakeward end of the federal breakwater. Transitions between typical sections and ends of typical sections also required additional detailing.

Since the Racine entrance breakwater was connected to the shore by approximately one-half mile of the north breakwater, construction procedures for the north breakwater reinforcement were critical to the construction of the entrance breakwater regarding placement procedures, tolerances, stone gradations, and weather. Though a majority of the contractors who bid the Racine project, including the ultimate successful bidder, were marine contractors, it was obvious that the placement of the core and armor stone by dumping from land-based trucks would present a major cost saving. Therefore, the specifications for the north breakwater incorporated a temporary haul road in the north breakwater reinforcement section to allow access to the entrance breakwater for direct land based construction. The feasibility of this haul road was a major factor in contractors' bid prices.

Tolerances employed in the specifications and construction of the breakwater were similar to those typical of a Corps of Engineers' rubblemound structure design. However, specifications were written to address the mass placement operation which was anticipated to be used for this project. Items addressed were requirements for placing the first two layers of armor stone, such that the underlying core stone layer not be disturbed, and a tolerance range for the outside slope of the berm structure indicating an angle of repose from 1.2 H:1V to 1.5 H:1V. (See Figure 3).

Measurement and payment sections were patterned after those typical for this type of construction on the Great Lakes. Measurements were made in tons (2,000 lbs/ton) by displacement of barges or from weight slips at certified scales. Payment was to be made on the basis of weight properly placed with the criteria for placement being determined on the basis of daily check surveys at 50 ft (15 m) or closer intervals, to the tolerances specified.

From the conceptual stage through bidding stage of the project, it was assumed that the placement operations for both core stone and armor stone could consist of a simple dump-and-push procedure by the contractor's haul trucks and bulldozers/backhoes. However, during the bid process it became obvious that the designers needed to acquaint bidders with the simplicity of the placement operation. Bidding contractors expressed doubt regarding whether the operation could be performed as simply as was described. Bidders who were given the opportunity to contact contractors who had actually built similar breakwaters gained confidence in the concept of this operation as a mass material procedure. In retrospect, a more concentrated effort in convincing and educating these contractors regarding the placement procedure may have been fruitful.

The bidding process also indicated several consequences of the virtual "selection" of the Vulcan Materials Racine quarry as the source for armor and core stone for the berm breakwater, due to use of stone sizes from this quarry in model testing. First, the quarry enjoyed a built-in advantage in negotiating with bidding contractors, frustrating the competitive market of a typical bid project. During the bidding phase, it was discovered that an alternative and economical source of core (and possibly also armor) stone was available, at some distance from the project. However, utilization of this stone source was not practical for bidding contractors, under the bid conditions existing. A second factor complicating the original assumptions regarding stone materials was that the quarry still placed traditional premium pricing on the larger, armor stone portion of its gradation. Therefore, despite the test blasting and planning of the modeling study to economically use the entire quarry gradation production, it was still cost effective to limit the volume of armor stone placement.

CONSTRUCTION MANAGEMENT

Unit price contracts were issued, typical for this type of construction and for the contractors who bid the work. A tenet of a unit price contract is that the contractor assumes the risk of determining a unit price which will be competitive, yet assures him of a proper return on his effort, while the owner assumes the risk of quantity. Therefore, it is important for the owner to be aware of the quantity conditions in a timely manner. In this project, the owner assumed a construction manager position, and performed the contract administration functions such as review of pay requests, measurement of quantities, direction of the contractor's work schedule, procurement and payment for materials, and review and acceptance of completed facilities. In several instances, the documentation of pre-construction conditions and the progress payment surveys were not completed in a timely fashion, contributing to some confusion. Specific instances of this problem include cases where a storm occurred before check surveys were completed, moving materials from a completed section which had been paid for on the basis of weight tickets, resulting in extra payments by the owner for replacement or reshaping of sections for which payment had already been made. An additional problem in payment occurred due to the tolerance given in the specification for the lakeward profile of the berm, which was intended to be built to specified width, and have a lakeward profile at the quasi-stable angle of repose for the armor. The tolerance of 1.2H:1V to 1.5H:1V listed, however, proved confusing for the contractor and owner in establishing the allowable pay section.

Material inspections were conducted at both the quarry and job site. Significant effort was undertaken at the quarry to attempt to meet the specified gradations and to specifically limit the amount of fine material in the armor stone. Materials which were passed through the quarry grizzly were segregated satisfactorily, however, during material dumping operations some stone was broken into smaller pieces, resulting in additional fines.

Since the Vulcan quarry at Racine is basically an aggregate quarry, the production of armor stone for the Racine Harbor project was a learning experience for this quarry operation, and required that the quarry invest in and set up equipment that was both expensive and whose use was unfamiliar to the operators. In addition, the test blast study which was undertaken for the project did not indicate the limited extent of competent Racine dolomite that existed in the quarry. Upon opening new faces of the quarry, the portion of quarry production smaller than the armor stone range increased substantially. Thus, in order to produce the needed volume of armor stone, the quarry was left with a large volume of smaller material which could not be used in the project. A larger test blasting operation or exploratory drilling may have indicated the rather limited extent of the more competent materials.

Construction of the berm breakwater proceeded from the end of the north breakwater, by first placing the core material (0 to 300 lbs) to an elevation permitting the use of the core as a roadway from which to place the armor stone (300 to 8,000 lbs). Armor stone was delivered by truck, and the berm constructed mainly by pushing with a bulldozer. The crest of the breakwater was completed using a large backhoe working toward the shore on completed crest sections (see Photograph 1).

Bulldozers working from the surface of the core material placing armor stone became a source of contamination, as fines from the core material were being worked into the armor stone. It was determined that contamination was occurring in a layer approximately 1 to 2 ft in thickness and was not occurring throughout the entire depth of the armor stone. Bulldozer operations were modified, limiting the surface over which they worked and subsequently reducing the amount of contamination. Equipment working over the surface of the armor stone and the handling and movement of armor stone units also produced a limited amount of breakage and contamination.

PHOTO 1. Berm breakwater construction. Crawler mounted backhoe moving stone into place on berm, with truck traffic on core stone.

Despite the difficulties noted above, the contractor made good progress in construction of the berm breakwater. North breakwater construction began in April and the entrance breakwater was complete, and accepted by the County, in November, 1986. Sections of in-place armor stone were measured, and were found to be within the specified gradation. Photograph 2 shows the completed berm in November. Note the substantial armor stone bench on the lakeward side of the breakwater crest.

PHOTO 2. Berm breakwater at completion. Note large bench at +8 ft., LWD in front of crest.

STORMS OF SPRING 1987

Two particularly severe storms occurred in 1987, on February 8th and 9th and March 8th, 9th and 10th. Due to abnormally warm winter conditions, no significant ice was present on the lake during these storms, allowing waves to directly impact on the entrance breakwater. The February storm caused major damage in the Chicago area, especially due to storm surge effects, but was not noted to be as extreme in the Racine area. No surveys were made of the breakwater after the February storm, before the storm of early March.

The March storm produced water level and wave conditions at Racine that were generally regarded as the worst in memory. The storm was characterized by sustained high winds over the longest fetch (north-northeast) of Lake Michigan exposed at Racine. Winds at Milwaukee were at or above 25 knots for approximately 24 h, with an approach direction of N 20° E to N 30° E.

Water levels at Racine were subject to a substantial storm surge due to the March storm. Interpolation of stage recorder data at Milwaukee and Chicago indicates that the storm produced water levels above elevations +4.5 ft, LWD (the project design water level) for approximately 20 h during the height of the storm, and above elevation +5.5 ft for approximately 6 h. The peak water level at Racine was estimated to be approximately +5.8 ft, 1.3 ft (0.4 m) above design water level. This peak water level occurred during the height of the storm and so probably coincided with the time of maximum wave action.

No wave height measurements were made during the storm, other than visual observations of wave heights which indicated that wave heights probably typically exceeded 10 ft. Very large spray plumes could be seen from shore from wave action on the entrance breakwater. A hindcast performed using Milwaukee storm wind data yielded predicted deep-water waves with peak H_S= 18.0 ft (5.5 m), T_S= 11.2 sec, from the north-northeast. A combined shoaling/refraction analysis, assuming straight and parallel bottom contours yielded predicted significant wave heights at the breakwater of 13.5 ft to 14.5 ft (4.1-4.4 m), with peak periods of 11 s. This analysis indicates that the storm of March 8th, 9th and 10th produced wave conditions approximating the structure design storm, with water levels up to 1.3 ft (0.4 m) higher than design water conditions.

Visual, underwater and survey assessments of the entrance breakwater were performed soon after the March storm. Photograph 3 shows the breakwater immediately after the storm. Significant observations on the condition of the breakwater are summarized as follows:

PHOTO 3. Breakwater after March storm. Note movement of armorstone bench.

- The 40-ft (12.2 m) wide berm originally placed at elevation +8 ft had been reshaped so that it was generally below water (elevation +3.5 ft) along the structure trunk. The berm surrounding the structure head was not as significantly reshaped.

- Significant numbers of rounded cobbles (6-in to 18-in diameter) were present near and above the waterline, concentrated at the western end of the breakwater, indicating breakage and reworking of some of the armor stone pieces. The majority of the cobbles seemed to be composed of the lighter-colored Waukesha dolomite.

- Some of the berm material appeared to have been moved up against the breakwater crest by wave action, forming a wider-appearing crest, with a fairly steep drop to waterline.

- A diving survey indicated that the mass of the berm was intact below water level, sloping lakeward at 6H:1V to 10H:1V. Pockets of rounded cobbles were noted below water along the breakwater trunk, with the great majority in water depths less than 5 ft (1.5 m) (-1.5 ft, LWD).

- No evidence of substantial overtopping of the breakwater was noted. The back slope of the breakwater crest appeared unaffected.

A subsequent survey was conducted by Racine County and indicated that despite the fairly dramatic change in above-water appearance, the entrance breakwater appeared to have behaved similarly to the model tests with respect to berm reshaping. Survey cross sections and profiles from the final modeling study are presented in Figures 4 and 5.

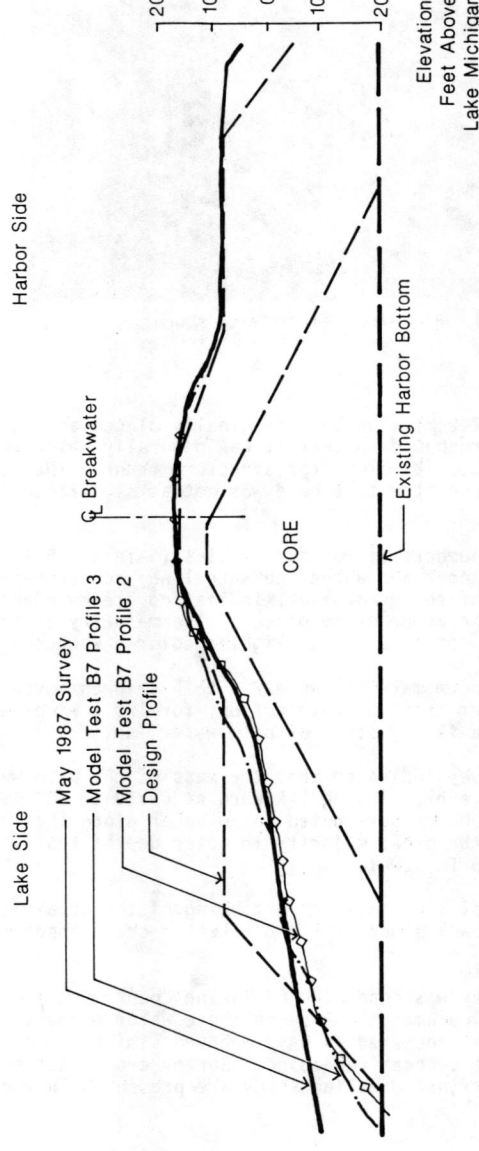

FIGURE 4 • BERM BREAKWATER SECTION NEAR HEAD OF BREAKWATER

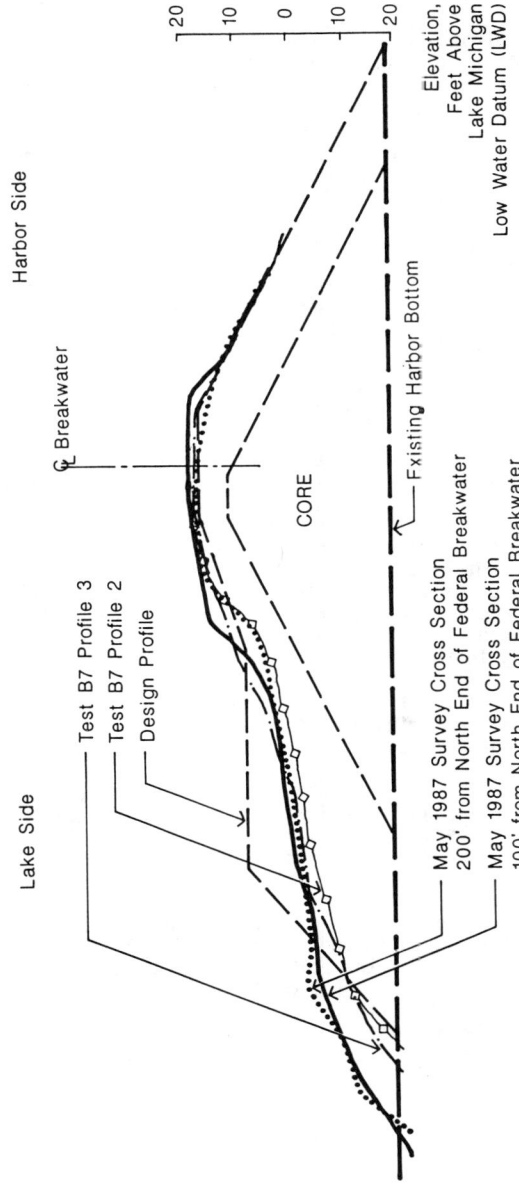

FIGURE 5 • MODEL AND PROTOTYPE SECTIONS AT TWO LOCATIONS IN BREAKWATER TRUNK

Substantial discussion regarding the condition of the breakwater ensued, regarding wave and water level conditions and the apparent breakage of some of the breakwater armor. However, the overall response of the breakwater suggested that the design modeling accurately simulated the major breakwater response. Review of in-place armor on the breakwater crest suggested that the smallest fraction of the armor (300 lbs through 1,000 lbs) (0.13-.45 t) may have been the source for most of the cobble material noted, and that substantial progressive failure of other armor units was not occurring. Following this line of reasoning, a reparative construction program was conducted, to provide a lakeward profile which approximated the profile which would have been called for if higher design water levels were used. This construction program consisted of regrading the crest armor and placement of additional armor on the upper portions of the lakeward berm, as indicated in Figure 6. The already-defined berm armor stone was used in this work, which was completed in June, 1987. No significant storms have occurred since placement.

CONCLUSIONS

The experience in building the Racine breakwater and observations of the effects of severe storms suggest the following conclusions:

- Efforts to take advantage of the competitive market for stone materials should not pre-suppose that local sources will be most economical. Utilization of physical modeling requires that the armor stone gradations be determined substantially before bidding documents are prepared. To avoid "locking in" on a particular source, the following approaches could be investigated in the future:

  - Pre-bidding for the stone supply source, possibly including the owner acquiring an option or issuance of a letter of intent for purchase of the armor for the project.

  - Issuance of bidding documents with alternative bid sections, developed through modeling of alternative gradations. At a minimum, a berm-type breakwater and a conventional 2-layer armor breakwater could be designed and bid as alternate sections.

- Criteria for stone quality important in berm breakwater construction may need further development. In particular, criteria for assessing the potential for breakage in the more-mobile berm breakwater reshaping process may need to be developed. It is important to note that while use of a berm concept may substantially relax gradation requirements for successful breakwater construction, stone quality requirements may be even more important.

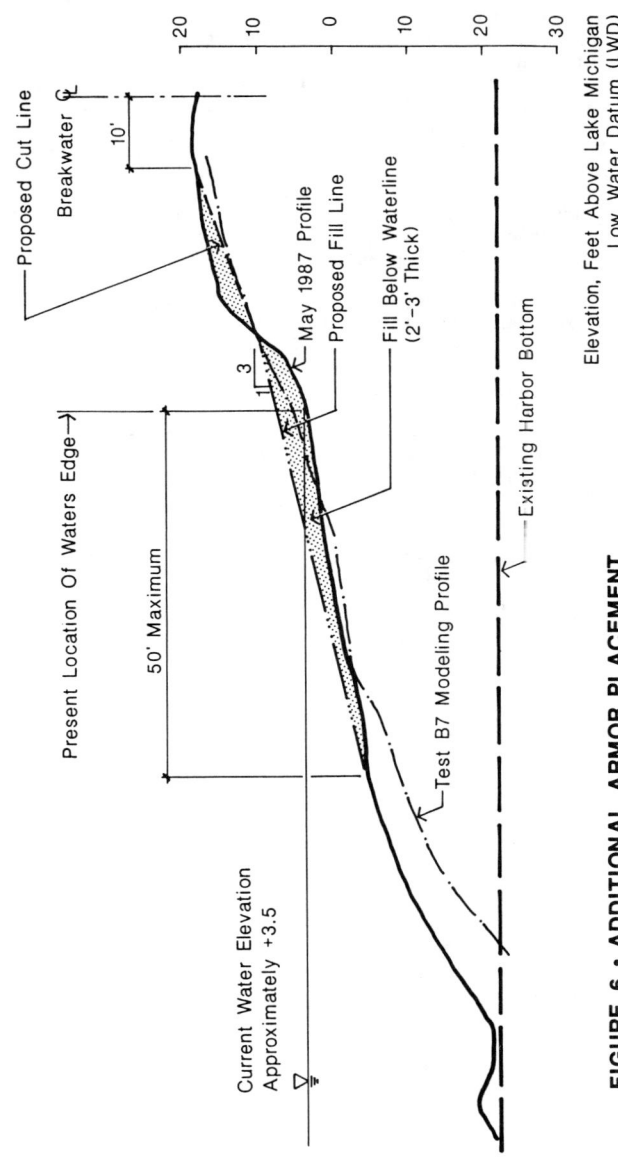

FIGURE 6 • ADDITIONAL ARMOR PLACEMENT

- A substantial program of educating the bidding contractors may pay off handsomely in more competitive bids. Special attention needs to be paid to allow land-based construction wherever it appears advantageous. A small demonstration construction project may even be justified on the largest projects, to give contractors a direct "feel" for berm construction conditions.

- Despite test blasting analyses, a substantial "learning curve" may be necessary for the quarry operation to consistently provide the gradation of material needed. In addition, commerical quarry operators may still attach a traditional price premium on the larger-sized stone materials, despite project planning to utilize the full gradation of quarry production.

- Investigation of the sensitivity of berm breakwater designs to design storm exceedance conditions as part of the modeling processes would be helpful in project planning.

ACKNOWLEDGEMENTS

The contributions of Cy Ingrahm, John Spooner and Larry Ryan are gratefully acknowledged. Warzyn Engineering encouraged and supported the preparation of this paper.

SYMBOLS AND DEFINITIONS

H_s — Significant Wave Height

T — Wave Period

W_{50} — Weight of 50 percent size of armor gradation

D_{n50} — Nominal diameter of armor unit of weight= W_{50}, taken as $(\rho_s/\rho_w)^{0.333}$

Δ — Relative mass density of stone, defined as $\rho_s/\rho_w - 1$

ρ_s — Mass density of armor stone

ρ_w — Mass density of water

REFERENCES

1. Hall, K and Anglin, D, 1985. Racine Harbor - Hydraulic Model Investigations. Report to Warzyn Engineering by W.F. Baird & Associates, Ottawa

2. Resio, D.E. and Vincent, C.L., Design Wave Information for the Great Lakes, Report 3, Lake Michigan, 1976. Hydraulic Laboratory, U.S. Army Corps of Engineers Waterways Experiment Station

3. Van der Meer, J.W. and Pilarczyk, K.W. 1986. Dynamic Stability of rock slopes and gravel beaches, Proceedings, 20th International Conference on Coastal Engineering, Taipei

4. Ziolkowski, L. and Ryan, L.W., 1987. Harbor Rehabilitation and Recreational Development in Racine, Wisconsin, Proceedings Coastal Zone 87, ASCE, New York

BERM TYPE ARMOR PROTECTION FOR A RUNWAY EXTENSION AT UNALASKA, ALASKA

by

Charles I. Rauw

Abstract

A proposed runway extension located on the Aleutian Islands was designed to withstand forces of Bering Sea storm generated waves using a unique armor protection concept. The proposed runway was to extend 2,600 feet seaward into the Bering Sea with water depths approaching 60 feet. The design storm had a significant wave height of approximately 30 feet at the head of the structure. Constraints on the size of local quarry materials led to alternative armor protection considerations of on- and off-site casting of concrete armor units, importing large armor stones via barge and a concrete caisson type protective retention wall on top of a rubble structure. These alternatives were discarded in favor of a unique armor design employing the concept of a rock berm which would allow utilization of 100 percent of quarry production in constructing the runway extension and armor protection. The berm concept of armor protection has since been applied to other rubble mound structures, including breakwaters and seawalls. These applications have resulted in lower construction bids, compared to conventional armor protection schemes. A brief historical perspective is presented as well as a discussion of advantages and disadvantages of the berm concept over conventional armor layer designs.

Résumé

Le prolongement proposé d'une piste d'attérissage située aux îles Aléoutiennes a été conçu de manière à résister aux forces exercées par les vagues soulevées par les tempêtes en mer de Béring grâce à un concept unique de carapace de protection. La piste proposée devait être prolongée de 2600 pieds vers le large dans la mer de Béring par des profondeurs d'eau approchant 60 pieds. La hauteur significative des vagues soulevées par la tempête nominale était d'approximativement 30 pieds au musoir de l'ouvrage. Des contraintes quant aux dimensions des matériaux extraits des carrières locales obligeaient à prendre en considération, pour la carapace de protection, des solutions de replacement comme le moulage sur place et ailleurs d'unités de carapace en béton, l'importation par chalands de grosses pierres de carapace et la construction d'un mur de rétention protecteur de type caisson en béton au sommet d'un ouvrage en enrochement. Ces solutions ont été rejetées au profit d'une conception unique de carapace faisant intervenir le concept d'une risberme de roches qui permettrait d'utiliser à 100 % la production de la carrière pour la construction du prolongement de la piste et de la carapace de protection. Le concept de risberme pour les carapaces de protection a été appliqué depuis à d'autres ouvrages en enrochements, dont des brise-lames et brise-mers. Ces applications on entraîné des coûts de soumission inférieurs pour la construction, comparativement aux méthodes classiques de construction de carapaces de protection. L'étude presente une brève perspective historique ainsi qu'une discussion des avantages et inconvénients du concept de risberme par rapport aux conceptions classiques avec couche de carapace.

BERM TYPE ARMOR PROTECTION FOR A RUNWAY
EXTENSION AT UNALASKA, ALASKA

By

Charles I. Rauw, Senior Coastal Engineer

Winzler & Kelly, Consulting Engineers, San Francisco, CA

Abstract

A proposed runway extension located on the Aleutian Islands was designed to withstand forces of Bering Sea storm generated waves using a unique armor protection concept. The proposed runway was to extend 2,600 feet seaward into the Bering Sea with water depths approaching 60 feet. The design storm had a significant wave height of approximately 30 feet at the head of the structure. Constraints on the size of local quarry materials led to alternative armor protection considerations of on- and off-site casting of concrete armor units, importing large armor stones via barge and a concrete caisson type protective retention wall on top of a rubble structure. These alternatives were discarded in favor of a unique armor design employing the concept of a rock berm which would allow utilization of 100 percent of quarry production in constructing the runway extension and armor protection. The berm concept of armor protection has since been applied to other rubble mound structures, including breakwaters and seawalls. These applications have resulted in lower construction bids, compared to conventional armor protection schemes. A brief historical perspective is presented as well as a discussion of advantages and disadvantages of the berm concept over conventional armor layer designs.

Introduction

The berm concept of armor protection is simply described as placement of a pre-determined volume of armor stone into a prism on the exposed surface of a structure or slope to be protected from wave action (Fig. 1). The critical design parameters of the berm are the gradation of the armor stone and the width of the berm. The crest elevation of the berm is generally related to the elevation that land-based construction equipment can safely operate at during normal wave action. The outboard slope of the berm is placed at the natural angle of repose of the armor stone gradation. This allows

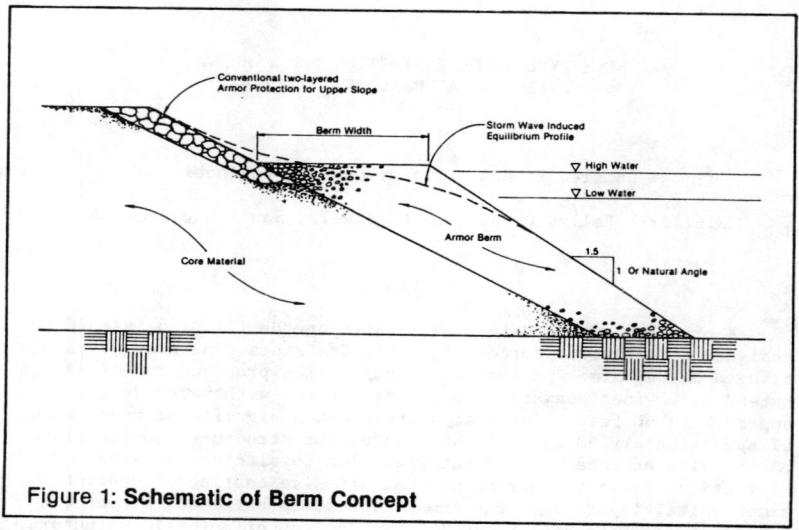

Figure 1: **Schematic of Berm Concept**

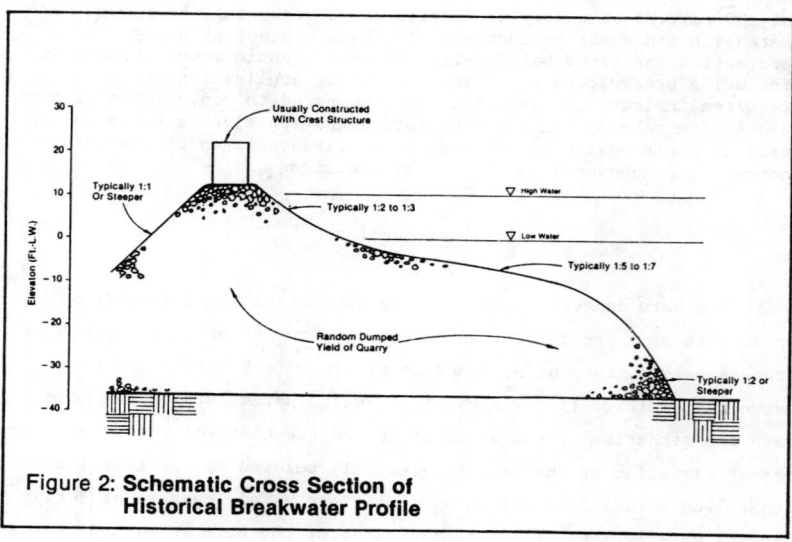

Figure 2: **Schematic Cross Section of Historical Breakwater Profile**

the berm to be constructed using only dumping or dumping and pushing operations. Additional slope protection, if required at elevations higher than the berm, can be constructed "in the dry", using conventional designs and construction techniques.

After construction, storm wave action reshapes the berm slope until an equilibrium profile is obtained, as shown schematically in Figure 1. The ability to obtain an equilibrium profile without damage to the structure is primarily dependent on berm width, once the material gradation and design storm characteristics have been specified. A stable berm width is determined from hydraulic model testing. During the reshaping process considerable stone movement occurs in the surface layer of the berm. This movement consolidates the outer layer and creates a natural sorting and nesting of stones. This process develops a high degree of interlocking in the outer layer of stones, which increases stability and resistance to wave action. In addition to this "self-armoring" process, the berm concept also derives stability from its relatively high porosity. This porosity allows waves to dissipate energy within a large matrix of stones, thus reducing hydraulic forces that individual armor stones need to resist in conventional two-layer designs with relatively impermeable filter layers.

The gradation of stone specified in the berm can be developed to best utilize available local material. The performance of various gradations of armor stone is established in hydraulic model tests. Gradations can be wide or narrow; however, model test results have indicated a wider gradation will produce greater interlocking in the surface armor layer. The average stone weight of armor gradations used in the berm concept has been considerably smaller than that required by conventional two-layered designs.

The berm concept is not a new concept, as a study of historical rubble mound structures clearly demonstrates. The berm concept reflects quite accurately the construction techniques and behavior of breakwaters constructed in the early 1800s. Baird (1983) reviewed technical literature on historical breakwaters to obtain an understanding of design procedures, construction techniques, and performance. Baird reported that many breakwaters constructed in the early 1800s (Kingstown, Portrush, Cherbourg, Holyhead and Plymouth) were constructed by using all material produced at a quarry and dumping the rock at the breakwater site to achieve a specified design cross section. The design cross section was usually based on observation of stable profiles of other rubble mound structures. Typically, these breakwaters had profiles that resembled the schematic section shown in Figure 2. The "S-shaped" slope of the seaward berm was dictated by wave action and gradation of dumped rock. Repair of these structures was simply a matter of dumping additional stone.

In general, conventional armor protection uses a layered design requiring careful placement of specified thicknesses and gradations of quarry stone or armor units. Advances in construction equipment capabilities and techniques made the layered design attractive as it required less volume of material. However, as structures were built in deeper and deeper water, two factors began to play an important role in the constructability of rubble mound structures.

First, deeper water, if located in areas exposed to severe wave attack, will generally result in larger design wave loading. This requires larger armor stones or armor units on the exposed face or flatter slopes to achieve stability in the armor layer. Quarry operations geared to maximize production of large armor stones can be characterized as a slow "mining" process with considerable waste. In addition, there is ultimately a maximum size of armor stone that can be produced from a given quarry. Casting concrete armor units, used where available stone size is inadequate, is a slow and expensive process. Flatter slopes, used to achieve stability in the armor layer, result in larger volume of quarry material for the core, filter and armor layers.

Second, due to the geometric shape of a rubble mound structure, the percentage volume of rock in the armor layer, compared to the total volume of rock in the structure, is reduced when the structure is sited in deeper water. Thus the demand for specified gradations of material from a quarry changes as structures are sited in deeper water depths. These factors are important considerations when examining the applicability of a conventional layered armor protection design.

Project-Specific Application

General

As an example application of the berm concept of armor protection, an overview of wave protection design for a proposed offshore rockfill runway extension at Unalaska, Alaska, is presented. A thorough discussion of this project was presented in Hall et al. (1983). The examination of the berm concept alternative was a result of local quarry sources' inability to easily and economically yield the size and volume of large armor stones required in a conventional layered armor protection scheme.

Site Description

The town of Unalaska and the adjacent Port of Dutch Harbor form an Aleutian island community of approximately 2,300 people located approximately 790 statute miles from Anchorage, Alaska (Figure 3). Unalaska and Dutch Harbor form a major fishing, stop-over and refueling port for sea-lift operations serving other Bering Sea communities, western and northwestern Alaska and the North Slope.

The existing airport at Unalaska was constructed during World War II and is approximately 4,000 feet long by 100 feet wide (Figure 4). The proposed widening, resurfacing and extention of the runway would allow operation of 737/727-class aircraft and link

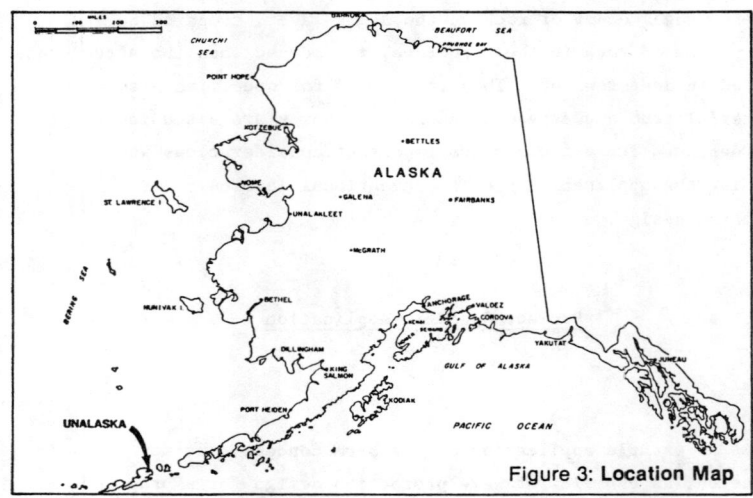

Figure 3: **Location Map**

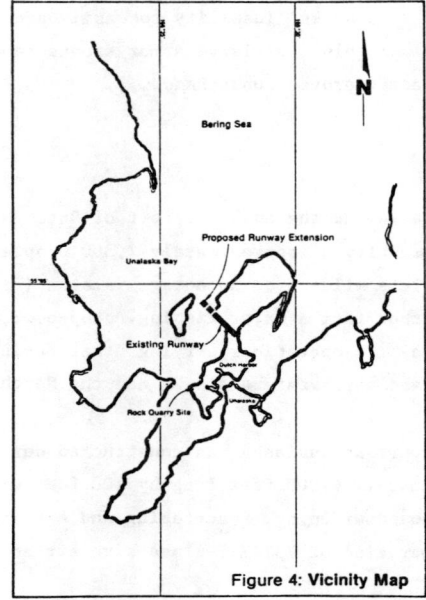

Figure 4: **Vicinity Map**

Unalaska with the existing jet system operating throughout Alaska. Airspace limitations and aircraft requirements led to the recommendation of extending the existing runway 2,000 feet into Unalaska Bay, where water depths approached 60 feet. This location has a direct exposure to Bering Sea storm waves that have little apparent attenuation from deep water to the site.

Design Criteria

A comprehensive feasibility study for upgrading the Unalaska airport examined site-specific design criteria (Dames and Moore, 1980). These studies were augmented during final design by field measurements, analyses and hydraulic model investigations. For the purposes of armor protection design, storm waves, tsunamis, earthquakes, storm surge and quarry yield were examined. The two most restrictive criteria for armor protection design were design storm wave conditions and projected quarry yield. Storm wave conditions were developed based on a 20-year historical hindcast calibrated with a storm wave measurement program. Results yielded a range of significant wave heights from 20 to 27 feet at the 95% confidence limit for an average recurrence interval of 100 years. Characteristic wave periods ranged from 10 to 12 seconds. A projected quarry yield was developed from results of a test blast at a selected quarry site (Figure 4). The following summarizes the quarry yield estimates, based on two assumed production rates.

Stone Size (tons)	Maximum Size Production (%)	Efficient Volume Production (%)
Greater than 24	-	-
16-24	3	1
6-16	13	10
1- 6	45	42
1	39	47

Design Alternatives

The basic structure of the proposed runway extension was a rock mound. The width, length and elevation of the structure was primarily set by airspace and aircraft requirements. The alternatives examined involved variations of wave protection provided around the exposed perimeter of the structure. These concepts included conventional embankment structures, pile- and caisson-supported runways, concrete structures and rock and concrete armor unit rubble mound structures. Based on the results of quarry yield investigations, armor stone was limited to a maximum size of 20-24 tons. Rubble mound structures were found to provide the greatest cost advantage of the alternatives considered at the feasibility level. Using conventional layered armor protection design and stability criteria recommended in the Shore Protection Manual (U.S. Army, 1977), several variations of rubble mound structures were examined.

During final design, hydraulic model tests were conducted to determine stability and response of the armor protection designs. The alternative selected for hydraulic model investigations was a benched slope rubble mound design. This concept embodies traditional armor slope design coupled with a relatively flat bench slope that causes large waves to break and dissipate energy prior to reaching the runway embankment slope. In an attempt to minimize or eliminate marine construction requirements, two separate designs were developed (Figures 5A & 5B). The berm concept evolved as a result of the model testing and site constraints discussed in the following section.

Hydraulic Model Tests

A series of three-dimensional hydraulic model tests were conducted at the Hydraulic Laboratory of the National Research Council in Ottawa, Canada, on the design concepts shown in Figures 5A and 5B. The tests were designed and run by W.F. Baird and

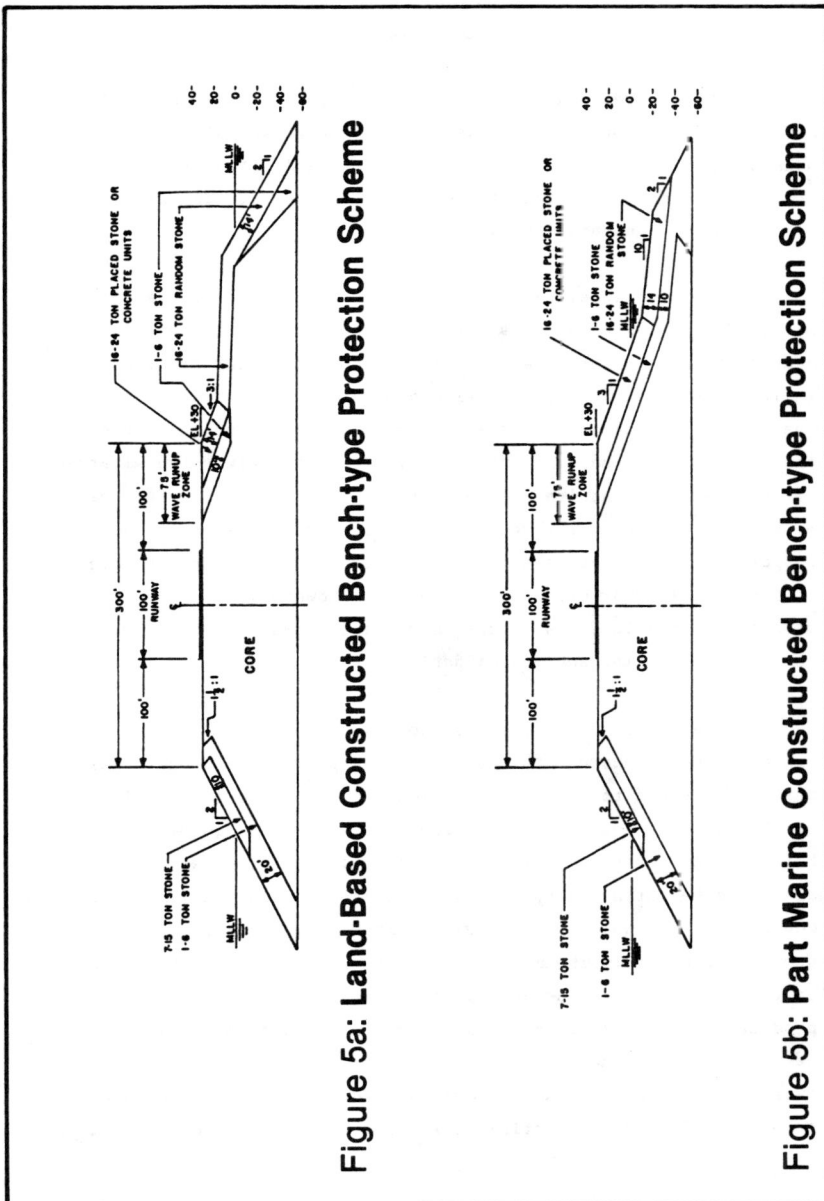

Figure 5a: **Land-Based Constructed Bench-type Protection Scheme**

Figure 5b: **Part Marine Constructed Bench-type Protection Scheme**

Associates, Coastal Engineers, Ltd., Ottawa, Canada, at geometric scales of 1:44 and 1:50. All tests were run using irregular waves having a spectra that simulated storm wave conditions measured at the site and corresponding to the design significant wave height and period characteristics. Materials used to construct the models were hand-picked angular crushed rock with a specific gravity ranging from 2.5 to 2.65. Details of the complete model testing program were presented in Hall (1982).

The initial model testing of the two bench concepts resulted in severe damage to the armor stone protection and significant overtopping. However, the resulting post-storm model profiles indicated that an S-shaped profile similar to that shown in Figure 2 was stable. A third alternative section was proposed that resembled the post-storm S-shaped profiles achieved by the first two alternatives. This alternative design section, shown in Figure 6A, was modeled and tested. The model results indicated that this section could survive the design storm event, although considerable movement of individual armor stones was observed. However, this movement of stones resulted in a post storm increase in armor layer stability due to interlocking and consolidation.

The berm concept evolved as the next step in the design process. One concern with the bench concept was the need to use marine equipment to construct portions of the slope (core, filter and armor) that could not be reached by land-based equipment. Mobilization costs and limited operational conditions at Unalaska did not make marine equipment use a desirable construction option. Concerns were also expressed over limitations of quarry source and quarry operations. The maximum size of quarry rock was estimated to range from 20 to 24 tons and, in order to achieve the maximum production of large stones, a slow "mining" quarry operation would be necessary. In addition, from a materials-balance standpoint, overproduction of the quarry would be necessary to provide the required volume and gradation of large armor rock. The combination

RUNWAY EXTENSION ARMOR PROTECTION 261

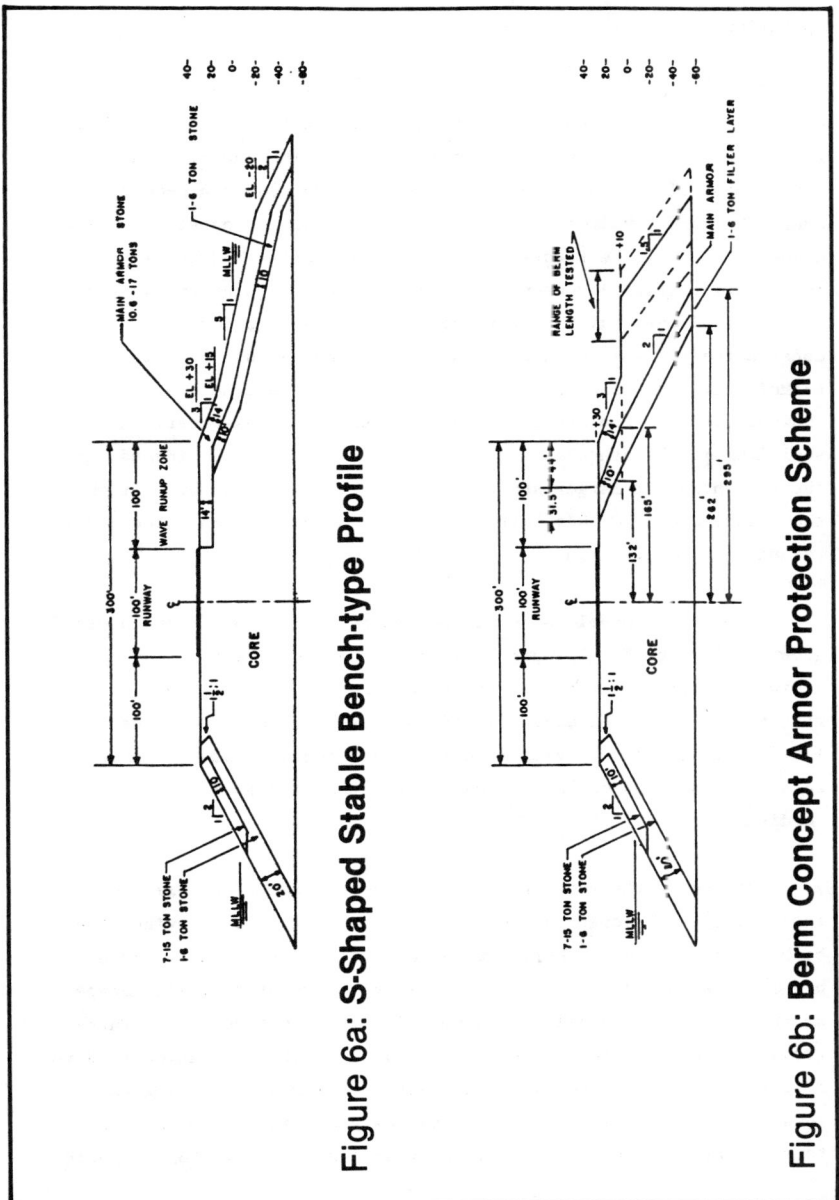

Figure 6a: S-Shaped Stable Bench-type Profile

Figure 6b: Berm Concept Armor Protection Scheme

of overproduction and slow quarrying techniques made quarry operations costly.

The berm concept was an attempt to minimize these two constraints. The tested bench design shown in Figure 6A exhibited a stable profile that was resistant to damage. The question was asked, that if a prism of material were end-dumped in a geometric shape that had a volume of armor material equal to the stable bench concept, would it be reworked to a similar profile and develop interlocking stability when subjected to the design storm event? The attractiveness of this berm concept was that it could be constructed entirely with land-based equipment using end-dump techniques. The upper portion of the armor protection could be finished off "in the dry", similar to capping a breakwater. In addition, if the gradation of the armor stone could be widened and the percentage of larger stones reduced, then in this application, the quarry could be "hit" harder, increasing production rates and reducing overproduction.

A series of model tests were conducted using the "berm concept" shown in Figure 6B. The performance of the design was measured against variations in the width of berm and specified stone gradation. The crest elevation of the berm was selected to be +10 ft.-mllw, based on estimated freeboard requirements for safe working conditions of land-based equipment during normal wave conditions.

Results of the model tests were encouraging. The berm was reworked by wave action during the design storm event, moving stones both up and down slope until sufficient interlocking had occurred to provide a stable slope. The stable slopes were of a similar shape as the equilibrium profile obtained with the bench concept. Three gradations of armor stone were examined; 1) a wide gradation (3.9 to 19.2 tons - ave. 9.0 tons); 2) a light narrow gradation (7.2 to 11.7 tons - ave. 9.1 tons); 3) a heavy narrow gradation (10.6 to 17.0 tons - ave. 13.3 tons). The wider gradation of armor stones showed a greater degree of interlocking than narrow gradations. The wider gradation showed more stability than the light narrow gradation and showed approximately the same degree of stability as the heavy narrow gradation.

The width of the horizontal berm was determined to be a critical design consideration from two perspectives. First, if the berm width was too short, interlocking of the armor stones did not occur prior to erosion of the berm and exposure of filter and core material. Second, as the berm width was shortened, run-up increased and the stability of armor stone on the upper slope decreased. Another factor attributed to increased run-up was the surface layer interlocking and consolidation that occurs during the initial period of profile readjustment. However, this is offset, when compared to conventional layered designs, by the wave energy dissipation that occurs within the large void volume of the berm concept.

In conclusion, the model tests indicated that the berm concept was as stable as the conventional layered bench design and could be constructed entirely using land-based equipment. This stability was achieved using a wide gradation of armor stones with an average weight 30 percent less than that required for the layered conventional bench design.

Discussion

The "development" of the berm concept of armor protection during the design process of the Unalaska runway extension is an excellent example of what can be learned or confirmed by examining prototype structures. The brief historical perspective presented in the Introduction serves to illustrate that the berm concept is not new. More importantly, it should demonstrate that the berm concept should not be viewed with skepticism as untried or unproven, since these prototype structures have been in existence for well over 100 years. The applicability of the berm concept for rubble mound structure design is related to several factors, one of the most important being the availability and production characteristics of local quarry materials. Design procedures for conventional layered armor protection schemes, such as those presented in the Shore Protection Manual (U.S. Army, 1984) do not make allowances for limitations in local quarry operations and materials or the efficient use of the yield of the quarry. In many instances this does not present a problem and conventional design procedures will provide an economical and stable structure. However, material sources should be carefully examined early in the design process and investigations conducted to establish production rates, size limitations and gradation capabilities.

The berm concept has received quite favorable reviews from contractors as well as designers. The construction technique is simple and does not require special equipment. The berm concept can be constructed using land-based equipment end-dumping material directly or using a dumping and pushing technique. The contractor does not have to maintain as close tolerances as with conventional layered designs and need not be concerned with underwater construction. The contractor is ultimately faced with placing a specified volume and gradation of material in a berm with a given width.

The berm concept has several advantages over conventional layered armor protection schemes. These advantages are present in the areas of materials, performance, constructability, and maintenance. Table 1 presents a comparison of attributes that make the berm concept an attractive alternative to conventional layered armor protection designs as well as disadvantages.

In addition to the Unalaska design, other projects have been designed using the berm concept (Baird, 1984). Table 2 lists some of the projects, the design wave conditions and the gradation of stone found to be stable using the berm concept. Table 2 also presents the stable armor stone gradation calculated for a conventional layered armor protection design using procedures

TABLE 1

COMPARISON OF BERM CONCEPT TO CONVENTIONAL LAYERED DESIGN

		Berm Concept	Conventional Layered Design
1.	Materials	o Uses available stone from quarry	o Uses slower "mining" techniques to obtain larger stones
		o Structure can be design to use 100% of quarry	o Quarry must operate to produce multiple specified gradations of armor, filter and core stones (may have considerable wastage)
		o Quarry production rates can be higher	o Alternately, large concrete armor units may be used. Units are expensive and casting, curing and handling is slow
		o Only two fractions of the quarry need to be specified (larger than x and smaller than x)	
		o Requires larger volumes in armor section	
2.	Performance	o Smaller stones required for same stability	o Hydraulic stability well established in prototype and model tests
		o Relatively higher porosity of berm allows waves to dissipate energy within berm and reduces up- and down-rush forces on exposed armor face	o Relatively impermeable filter and core layers result in larger up-and down-rush forces on armor layer

TABLE 1 (Con't)

	Berm Concept	Conventional Layered Design
	o Loss of surface armor stones does not expose structure core rapidly (considerable reserve armor material in berm)	o Loss of armor layer exposes unstable filter and core material (rapid failure of slope can occur)
	o Requires model testing to confirm design and establish berm width and armor gradation	
	o Stone durability an important consideration, since initial profile adjustment requires considerable motion	o Concrete armor units are not at present properly designed for wave-induced stresses (designed for hydraulic stability only)
3. Constructability	o Simple and fast construction techniques may be used (construct a specified width of berm at a specified elevation)	o Construction is slow due to careful placement and specified tolerances required in armor, filter and core materials
	o Only land-based equipment is required and construction is not as limited by environmental factors	o Underwater placement requires strict supervision
	o Construction tolerances can be reduced	o Deepwater structures may require expensive marine construction equipment and techniques. Marine construction is more sensitive to environmentally limited operating conditions

TABLE 1 (Con't)

	Berm Concept	Conventional Layered Design
4. Maintenance	o Maintenance is simple and performed by dumping more material and dressing the slope, if desired.	o Requires mobilizing specialized and costly equipment
		o Armor protection failure may result in considerable mixing of armor, filter and core material necessitating expensive excavating prior to repair or overbuilding to achieve a stable layered design

TABLE 2

BERM CONCEPT DESIGNS

LOCATION	WAVE CHARACTERISTICS	BERM CONCEPT STABLE STONE GRADATION (TONS)			CONVENTIONAL LAYERED DESIGN[1] STABLE STONE GRADATION (TONS)		
		MIN	MEAN	MAX	MIN	MEAN	MAX
alaska, Alaska	H_s = 32.8 ft Tpeak = 12 sec	3.9	9.0	19.0	55	74	93
lgavik Bay, Iceland	H_s = 19.7 ft Tpeak = 9.6 sec	1.7	3.0	7.0	12	16	20
droy, Newfoundland	H_{max} = 29.9 ft Tpeak = 12 sec (H_s estimated at 18 ft)	0.5	—	5.0	9	12	15
rth Bay, Ontario	H_s = 4.9 ft Tpeak = 6 sec	0.002 (4.4 lbs)	—	0.22 (440 lbs)	0.19 (370 lbs)	0.25 (494 lbs)	0.31 (617 lbs)

[1] Assume structure slope of 1:5, K_d = 2, w_r = 165 lb/ft^3

presented in the Shore Protection Manual (U.S. Army, 1977). The differences in required stone weights are considerable.

The largest disadvantage in considering the berm concept is the present lack of empirical data that can be used to develop design conditions. Each application will require some model testing to confirm stability and gradation specifications until analytical relationships are established or enough model tests have been made to develop empirical relationships. As more experience is obtained with prototype structures and model testing results the design process will become application oriented.

References

Baird, W.F. and Hall, K.R., "The Design of Armour Systems for Rubble Mound Breakwaters" Speciality Conference on Rubble Mound Breakwaters, ICE, London, England, May, 1983.

Baird, W.F., "Personal Communication", 1984.

Dames and Moore, "Site Design and Cost Studies for Feasibility, Assessment - Offshore Runway Extension at Unalaska Airport, Alaska", unpublished report for the State of Alaska, Department of Transportation, 1980.

Hall, K.R., Baird, W.F. and Rauw, C.I. "Development of a Wave Protection Scheme for a Proposed Offshore Runway Extension at Unalaska Airport, Alaska", Coastal Structures '83, ASCE, Arlington, Virginia, March. 1983.

Hall, K.R. and Baird, W.F., "Offshore Runway Extension of Unalaska Airport, Alaska - The Runway Extension from Wave Action Using a Hydraulic Model Study", unpublished report prepared for Dames and Moore, Los Angeles, Calif., May, 1982.

U.S. Army, Shore Protection Manual, Corps of Engineers, Coastal Engineering Research Center, 1977.

U.S. Army, Shore Protection Manual, Corps of Engineers, Coastal Engineering Research Center, 1984.

UNCONVENTIONAL RUBBLE-MOUND BREAKWATERS - CONCERNS

By
C.D. Anglin[1]
K.B. Dean[2]
D.H. Willis[2]

ABSTRACT

A Seminar on Unconventional Rubble-Mound Breakwaters was held in Ottawa, Canada, on the 15th and 16th of September 1987. This paper is an attempt to summarize the general discussion that was the final item on the programme. An unconventional rubble-mound breakwater is one in which the armour is quarried stone placed in any way other than the conventional two layers to protect against wave action. The following factors were discussed: design procedures; long-term stability; porosity and permeability; profile development; oblique wave attack and roundheads; scour; model testing; consolidation; stone durability; and construction. Recommendations were made concerning: field measurements; modelling of flow within the breakwater; the design and construction team; design storms; and improvements to conventional rubble-mound breakwater design.

RÉSUMÉ

Un séminaire sur les brise-lames en enrochements non classiques a été tenu à Ottawa, au Canada, les 15 et 16 septembre 1987. La présente communication tente de résumer la discussion générale qui constituait le dernier article au programme. Un brise-lames en enrochements non classique est un brise-lames dont la carapace consiste en enrochements de carrière, sauf de manière classique de deux couches, pour protéger contre les vagues. Les facteurs suivants ont été discutés: méthodes de conception, stabilité à long terme, porosité et perméabilité, évolution du profil, têtes rondes, affouillement, essais sur modèles, consolidation, durabilité des pierres et construction. Des recommandations ont été formulées concernant: les mesures sur le terrain, la modélisation de l'écoulement à l'intérieur des brise-lames, l'équipe de conception et de construction, les tempêtes nominales et les améliorations de la conception des brise-lames en enrochements classiques.

---

[1] Queen's University and W.F. Baird & Associates Limited. Kingston, CANADA
[2] Hydraulics Laboratory, National Research Council of Canada. Ottawa, Ontario, CANADA

SEMINAR ON UNCONVENTIONAL RUBBLE-MOUND BREAKWATERS

A seminar on unconventional rubble-mound breakwaters was convened on September 15th and 16th, 1987, by the American Society of Civil Engineers (ASCE), the Canadian Society for Civil Engineering (CSCE) and the National Research Council of Canada (NRCC). Within the context of this seminar, "unconventional" was understood to imply the use of natural stone or quarried rock, usually graded as opposed to being of uniform size, placed in any way other than the conventional two layers. As we discovered, this construction goes under a variety of names: berm, reef or mass armoured breakwaters. The seminar was held at the National Research Council of Canada, in Ottawa.

Twenty-two 'experts' (people with at least some experience with this type of breakwater) responded to the invitation to attend. Over the two days, they presented, listened to and discussed 13 technical papers covering the design, model testing and construction of this type of breakwater, see References. There was also a visit to the NRCC Hydraulics Laboratory to witness models of both conventional and unconventional rubble-mound breakwaters subject to wave attack. At the conclusion as Mr. Magoon summarized, it was apparent that this form of structure has been used successfully and that substantial cost savings have been achieved. However, more research is required before designs can be prepared without the support of developmental work that is likely to include the use of physical models.

Our present interest is with the final item on the programme - a general round-table discussion of concerns, research needs if you prefer. From the discussion of the individual presentations, we prepared the following list of subjects requiring attention:

Design Procedures - is it useful to develop formulae, similar to the Hudson-Iribarren formula for conventional breakwaters, or numerical models for the design of unconventional rubble-mound breakwaters?

Long-Term Stability - should these breakwaters be designed to reach a reasonably static equilibrium or should stone movement be allowed throughout the life of the structure?

Porosity and Permeability - is high permeability critical for stability and wave energy dissipation?

Profile Development - what are the characteristics of the stable profile which develops under wave action, and what parameters affect the profile?

Oblique Wave Attack and Roundheads - if the breakwater material can be moved by waves, what is the effect of transport along the trunk and away from the roundhead?

Scour - does unconventional construction also require unconventional techniques for preventing seabed scour at the toe?

Model Testing - if the designs are based on physical models, what are the limitations of these tests and how can this be expressed in quantitative terms?

Consolidation - how much consolidation of the mass of armour actually occurs and what are the consequences of the consolidation?

Stone Durability - is stone durability a particular concern for this type of structure and what criteria can be established?

Construction - what are the advantages and problems associated with construction specifications, methods and inspection for these structures.

The discussion began with consideration of these topics, which will be dealt with in subsequent sections of this paper. At the conclusion, each participant was asked to state what research he felt had the highest priority in the immediate future; the results of this exercise are summarized under "Recommendations".

DESIGN PROCEDURES

No set of design procedures resulted from the discussion at this conference. There was a question of whether in fact we have sufficient information on the dependent and independent parameters involved to develop models at this time. On the other hand, a number of engineering tools were recommended for designing berm breakwaters. Many of the participants had faith in the design process using hydraulic model studies. It was agreed that currently, this is the most reliable approach to breakwater design.

One numerical model was presented at this conference (van der Meer, 1987). This can be applied as a tool to investigate a variety of design scenarios. Further development in the numerical modelling of breakwater flow kinematics and the resulting behaviour is encouraged in the future.

LONG-TERM STABILITY

In the long term, berm breakwaters can be designed to be in either static or dynamic equilibrium. For breakwaters in dynamic equilibrium, maintenance of the armour layer may be required during the life of the breakwater for a number of reasons discussed below. The amount of maintenance will be dependant on the degradation rate of the structure.

Several factors influence the degradation rate of a structure designed for dynamic equilibrium. For example, stone movements may cause abrasion and fracturing of the stones in a particularly mobile structure where stone durability is insufficient. In addition, storms occurring at various water levels may result in a net "loss" of armour stone towards the toe of the structure, and the upper slope may eventually require maintenance to provide adequate stability during storms at high water levels.

A major factor in the degradation of any breakwater is the exceedence of the design event. In the case of the berm breakwater, the base of the breakwater will likely still be intact and remedial measures would entail replacement of armour above the water line. This maintenance would likely be considerably less costly than remedial measures for a conventional breakwater.

There was a lively discussion concerning the advantages and disadvantages of designing a structure which would be in either static or dynamic equilibrium. Some favoured the static equilibrium approach due to its safety and reliability, as well as concerns about stone durability. Others argued that often static equilibrium results in too costly a design and that the client may be prepared to accept some maintenance costs for a reduced initial investment. It is a site-specific decision for the engineer on how to balance initial and maintenance costs through the design procedure.

POROSITY AND PERMEABILITY

The porosity of berm breakwaters is in the range 40 - 45%. Not much change in this porosity has been observed before and after consolidation of these breakwaters by waves. The discussion of permeability and the porous nature of berm breakwaters raised some important concerns. There is in fact an essential need to understand the role of these parameters. One of the participants cited examples suggesting that we, as engineers, do not really understand much about the behaviour of berm breakwaters. The Plymouth breakwater has stood the test of time and it is considered to be impermeable. Another example is the mass of material left after the Sines breakwater failure; it was very impermeable and the slope of this material was also quite stable. These examples are contrasted with the generally accepted idea that high permeability must be achieved in the design and construction of berm breakwaters.

PROFILE DEVELOPMENT

As a general observation, the berm breakwater is a passive design, the basic idea being to adjust to wave attack. The development of a stable profile, static or dynamic, of a berm breakwater occurs due to wave action. Stones will move once wave conditions exceed a certain threshold. The profile develops in response to the local sea state, and further changes to the profile will only occur when the threshold, possibly now higher, is exceeded by different waves and water levels than those which formed the original profile. Further work is required to develop an understanding of the influence of various parameters on profile development, including:

- Stone size relative to wave conditions
- Stone gradation
- Stone density
- Stone shape

- Permeability, porosity of armour "layer"
- Water level variations

OBLIQUE WAVE ATTACK AND ROUNDHEADS

Roundheads are a particulary difficult portion of any breakwater to design. In the case of berm breakwaters, the possibility of stone transport around the roundheads can create special design problems and further research is necessary.

The design of roundheads must also take into account ship navigation. In rough weather, narrow harbour entrances can be treacherous and the use of a berm roundhead may result in a narrower entrance than a conventional or vertical walled structure.

It was the experience of a number of the participants that it is very difficult to merge a roundhead of a conventional type with a berm breakwater trunk. Hydraulic model results have indicated that displacement of the armour stone of the unconventional breakwater occurred next to the conventional roundhead.

Oblique wave attack is an important consideration in the case of a breakwater which is in dynamic equilibrium. In this case, armour stone can be moved along the structure in the same manner as sand is moved along a beach by oblique waves.

SCOUR

This topic was not discussed in detail. Scour might not be of importance to these unconventional structures, due to reduced wave reflections caused by the higher permeability of the berm structure resulting in lower shear stresses at the base of the structure.

FIELD MEASUREMENTS

There was concern among the participants regarding the general lack of field measurements available on unconventional breakwaters. Field measurements are important to evaluate the scale effects in physical modelling, as well as improving the general understanding of the behaviour of berm breakwaters.

Specific field measurements required are:

1. waves and water levels at the breakwater;

2. pore water pressures within the breakwater;

3. phreatic surface within the breakwater;

4. water velocities within the breakwater - may be difficult to reproduce in a model; and

5. profiles of the breakwater surface.

MODEL TESTING

All rubble-mound breakwater designs, both conventional and unconventional, are based on small scale model studies. The concerns raised about physical modelling are therefore not confined to berm breakwaters. These concerns were: true simulation of sea states in shallow water; model rock durability; and the modelling of wave propagation in porous media - permeability and air-water mixtures.

It was beyond the scope of this seminar to make recommendations with regard to the realistic simulation of sea states, but all agreed that more research should be done in this area.

Since the permeability of berm breakwaters is an important factor in their performance, the modelling of permability must also be important. It was noted that the simultaneous modelling of permeability and porosity was not possible. It is possible to model the steady-state permebililty by exaggerating the porosity. However, in this case, water storage is also exaggerated. A much less economical substitute, would be to scale the fluid properties as well as porosity, but for practical physical models, this is not possible. The permeability in the model then will be less than required by correct scaling, resulting in a conservative test of breakwater stability. Further research is required on the modelling of permeability under waves.

The action of air-water mixtures on prototype breakwaters may have an influence on the stability of armour units. It is possible that air inside the full-size armour layer voids is compressed by waves breaking, exerting significant forces on the surrounding armour. The modelling of this phenomenon has only been cursorily looked at and more investigation must be done in the future. The same applies to the modelling of stone durability, see below.

CONSOLIDATION

The consolidation of the armour units on a berm breakwater is a natural process occurring under wave attack. However, there was no general agreement among the participants on the amount of consolidation and its effect on the performance of the breakwater. Participants reported 0 to 20% consolidation in model tests. Consolidation makes the armour layer more stable due to the nesting of the stones and increases the shear strength of the berm but does not significantly decrease porosity or permeability. It is therefore probably not a serious concern.

ROCK DURABILITY

There was considerable discussion of the importance of rock durability in a berm breakwater, particularly one designed to be dynamically stable. Deterioration of the stones, due to abrasion and fracturing as the berm reshapes during storms, may have a significant influence on the performance of the structure. Current modelling technology does not allow us to model this process. Further work is required to develop designs which account for the degradation of the stones where this will occur in the prototype. In addition, work is required to develop specifications that ensure suitably durable stones are used for a given breakwater project.

The evaluation of the durability of the stone normally utilises standard geological engineering techniques. An important question to consider is whether or not these procedures produce results which can be considered indicative of the durability of stone under wave action.

CONSTRUCTION

It was generally agreed that there are significant advantages associated with the ease of construction of unconventional rubble-mound breakwaters, and that the chief problem encountered was simply the unfamiliarity of contractors with this type of construction. Given an adequate set of documents, any contractor familiar with conventional rubble-mound breakwaters could build an unconventional one. The most important factor was the contractor's understanding of the principles underlying the operation of the unconventional breakwater, i.e. permeability and profile development.

Of these, permeability was the chief concern. Temporary construction roads form areas of reduced permeability and should be carefully controlled. Roads perpendicular to the breakwater axis are less problem than those parallel to it, since they affect the profile only at discreet, fairly narrow, intervals. In any case, all construction roads within and below the zone of wave activity must be removed in order to ensure the correct functioning of the breakwater.

For the same reason, that is maintenance of permeability, the percentage of fines in the armour must not exceed specified limits.

In most unconventional rubble-mound breakwaters to date, the structure has been designed to maximize the yield from a specific quarry or to utilise readily available least-cost material. This puts the quarry owner in a very strong position, and leaves the structure owner and designer with responsibility for the supply of adequate material. It may therefore be preferable to build this type of breakwater under a turn-key contract, with design, testing, supply, construction and performance all the responsibility of the contractor; or to bid two designs, conventional and unconventional.

RECOMMENDATIONS

As noted above, the discussion finished with each of the participants answering the question, "What needs to be done next?" Somewhat surprisingly, there were fewer answers than participants:

1. The majority favoured field measurements, implicitly to answer questions about how well models of rubble-mound breakwaters represent nature. There were a number of suggestions, pore water pressures, velocities within the pores and stone

accelerations for example, but one of the most practical was simply to measure breakwater profiles before and after storms.

2. The modelling of flow within the breakwater was also mentioned explicitly. Not enough is known about permeability under oscillatory flow, nor about the effects of air trapped in the pores, to allow much confidence in our ability to model waves in porous media. Some simple laboratory tests could improve this level of confidence.

3. With physical modelling almost the only design tool available for unconventional rubble-mound breakwaters, it was felt that an hydraulics laboratory should form part of the team designing and constructing these structures. For example, it is essential that the laboratory be available to test variations in the stone size and gradation encountered during construction.

4. More thought must be given to the design event. European and Australian practice is for a 50 or 100 year storm, whereas U.S. practice in the Great Lakes is only 20 years, all based on statistical analysis. How do "deterministic" parameters, for example long-term trends in water level, fit into this probabilistic calculation? Do we know enough about the nonlinear response of these structures to settle on a single design event, even when we can agree on return period?

5. And finally we should not forget conventional rubble-mound construction, which in many cases remains more economical. The lessons learned with unconventional breakwaters must also be applied to the conventional type, perhaps through the development of more open concrete armour units and the use of more permeable breakwater cores.

ACKNOWLEDGEMENTS

The authors were only the listeners and note-takers during the discussion. We wish to thank our sources, those who took part in the seminar and discussions: John Ahrens, William Allsop, Bill Baird, Bill Bremner, Hans Burcharth, D.D. Davidson, Jeff Gilman, Chris Glodowski, Kevin Hall, Hans Hesen, Orville Magoon, Étienne Mansard, Jentsje van der Meer, Rob Montgomery, Yvon Morin, Chuck Rauw, Bob Richter, Otavio Sayao, Torben Sorensen, Garry Timco and Dave Turcke. We hope we have succeeded in representing their views.

We also wish to thank Ms. Heather Knudson, an unpaid secretarial student from Colonel By High School, who in the name of education deciphered our three very different scrawls to produce this paper.

REFERENCES

AHRENS, J.P.. 1987. Reef breakwater response to wave attack. This volume.

ALLSOP, N.W.H. and J.P. LATHEM. 1987. Rock armouring to unconventional breakwaters: the design implications of rock durability. This volume.

BAIRD, W.F. and K. WOODROW. 1987. The development of a design for a breakwater at Keflavik, Iceland. This volume.

BREMNER, W., B.A. HARPER and D.N. FOSTER. 1987. The design and construction of a mass armoured breakwater at Hay Point, Australia. This volume.

BURCHARTH, H.F. and P. FRIGAARD. 1987. On the stability of berm breakwater roundheads and trunk erosion in oblique waves. This volume.

GILMAN, J.. 1987. Performance of a berm roundhead in the St. George breakwater system. This volume.

HALL K.B.. 1987. Experimental and historical verification of the performance of naturally armouring breakwaters. This volume.

HESEN, J.C.G.. 1987. Artificial island protection off the coast of Malaga, Spain. Unpublished.

JUUL JENSEN, O. and T. SORENSEN. 1987. Hydraulic performance of berm breakwaters. This volume.

MANSARD, E.P.D.. 1987. Towards a better simulation of sea states for modelling of coastal structures. This volume.

van der MEER, J.W.. 1987. Application of a computational model on berm breakwater design. This volume.

MONTGOMERY, R.J., G.J. HOFMEISTER and W.F. BAIRD. 1987. Implementation and performance of berm breakwater design at Racine, Wisconsin. This volume.

RAUW, C.I.. 1987. Berm type armor protection for a runway extension at Unalaska, Alaska. This volume.

LIST OF PARTICIPANTS

PARTICIPANT	AFFILIATION
Mr. John Ahrens	Coastal Engineering Research Center U.S. Army Corps of Engineers Waterways Experiment Station P.O. Box 631 Vicksburg, MS 39180 U.S.A.
Mr. N.W.H. Allsop	Hydraulics Research Limited Howbery Park Wallingford, Oxfordshire, OX10 8BA England
Mr. W.F. Baird	W.F. Baird & Associates Ltd. Suite 150 38 Antares Drive Ottawa, Ontario K2E 7V2 Canada
Mr. W. Bremner	Blain, Bremner & Williams, Pty. Ltd. 47 Castlemaine Street Milton, Queensland 4064 Australia
Prof. H.F. Burcharth	Department of Marine Civil Engineering Aalborg Universitetscenter P.O. Box 159 DK-91 Aalborg, Denmark
Mr. D.D. Davidson	Coastal Engineering Research Center U.S. Army Corps of Engineers Waterways Experiment Station P.O. Box 631 Vicksburg, MS 39180 U.S.A.
Mr. Jeff Gilman	Peratrovich, Nottingham & Drage 1506 West 36th Avenue Suite 101 Anchorage, AK 99503 U.S.A.
Mr. C.W. Glodowski	Marine Directorate Public Works Canada Sir Charles Tupper Building Confederation Heights Ottawa, Ontario K1A 0M2 Canada

PARTICIPANT	AFFILIATION
Dr. K.R. Hall	Department of Civil Engineering Queen's University Ellis Hall Kingston, Ontario K7L 3N6 Canada
Mr. J.C.G. Hesen	Ballast Nedam Engineering B.V. P.O. Box 500 1180 BE Amstelveen The Netherlands
Mr. Orville T. Magoon	President American Shore and Beach Preservation Association P.O. Box 279 Middletown, CA 95461
Dr. E.P.D. Mansard	Hydraulics Laboratory National Research Council Canada Ottawa, Ontario K1A 0R6 Canada
Dr. Jentsje van der Meer	Harbours, Coasts and Offshore Technology Division Delft Hydraulics Laboratory P.O. Box 152 8300 AD, Emmeloord, The Netherlands
Mr. Robert J. Montgomery	Warzyn Engineering One Science Court University Research Park P.O. Box 5385 Madison, WI 53705 U.S.A.
Mr. Yvon Morin	Marine Directorate Public Works Canada Sir Charles Tupper Building Confederation Heights Ottawa, Ontario K1A 0M2 Canada
Mr. Charles Rauw	Connmass Consulting Civil Engineers 597 Center Avenue Suite 320 Martinez, CA 94553 U.S.A.
Mr. Robert Richter	Construction Consultant Journey's End Croton-on-Hudson, NY 10520 U.S.A.
Dr. O.J. Sayao	F.J. Reinders & Associates Canada, Ltd. P.O. Box 278 Brampton, Ontario L6V 2L1 Canada

PARTICIPANT	AFFILIATION
Mr. Torben Sorenson	Danish Hydraulic Institute Agern Alle 5 DK-2970 Horsholm, Copenhagen, Denmark
Dr. G.W. Timco	Hydraulics Laboratory National Research Council Canada Ottawa, Ontario K1A 0R6 Canada
Dr. D. Turcke	Department of Civil Engineering Queen's University Ellis Hall Kingston, Ontario K7L 3N6 Canada
Mr. D.H. Willis	Hydraulics Laboratory National Research Council Canada Ottawa, Ontario K1A 0R6 Canada

SUBJECT INDEX

Page number refers to first page of paper.

Airport runways, 250
Alaska, 250
Armor units, 41, 92, 104, 138, 147, 250
Australia, 147

Bering Sea, 219
Berms, 55, 73, 92, 219, 229, 250
Breakwaters, 21, 41, 104, 138, 219, 229

Coastal structures, 1
Computer models, 92
Construction methods, 138, 229

Design storms, 219, 250
Dynamic stability, 92

Erosion, 55

Great Lakes, 229

Hydraulic models, 104
Hydraulic performance, 73

Iceland, 138

Reefs, 21
Rock masses, 147
Rocks, 41
Rubble-mound breakwaters, 21, 55, 73, 147, 270

Sea state, 1
Simulation, 1
Slope stability, 41
State-of-the-art reviews, 270

Water depth, 92
Wave action, 55, 92, 219
Wave forces, 21
Wave generation, 1

AUTHOR INDEX
Page number refers to first page of paper.

Ahrens, John P., 21
Allsop, N. W. H., 41
Anglin, C. D., 270

Baird, W. F., 138
Baird, William F., 229
Bremner, W., 147
Burcharth, Hans F., 55

Dean, K. B., 270

Foster, D. N., 147
Frigaard, Peter, 55

Gilman, Jeffrey F., 219

Hall, Kevin R., 104
Harper, B. A., 147
Hofmeister, Gregory J., 229

Jensen, Ole Juul, 73

Latham, J. P., 41

Mansard, E. P. D., 1
Montgomery, Robert J., 229

Rauw, Charles I., 250

Sørensen, Torben, 73

van der Meer, J. W., 92

Willis, D. H., 270
Woodrow, K., 138

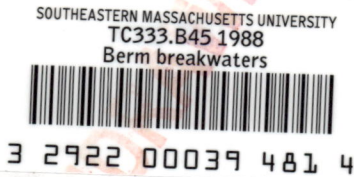

DATE DUE

NOV 2 5 2005			
OCT 3 1 2005			

Demco, Inc. 38-293